Error Coding Cookbook

Practical C/C++
Routines and Recipes for
Error Detection and Correction

Error Coding Cookbook

Practical C/C++ Routines and Recipes for Error Detection and Correction

C. Britton Rorabaugh

McGraw-Hill

New York San Francisco Washington, D.C. Auckland Bogotá
Caracas Lisbon London Madrid Mexico City Milan
Montreal New Delhi San Juan Singapore
Sydney Tokyo Toronto

McGraw-Hill

*A Division of The **McGraw·Hill** Companies*

hc   1 2 3 4 5 6 7 8 9 DOC/DOC 9 0 0 9 8 7 6

Library of Congress Cataloging-in-Publication Data
Rorabaugh, C. Britton.
   Error coding cookbook : practical C/C++ routines and recipes for error
detection and correction / by C. Britton Rorabaugh.
     p.   cm.
   Includes index.
   ISBN 0-07-911720-1 (h)
   1. C (Computer program language)   2. Debugging in computer
science. I. Title.
QA76.73.C15R67   1995
005.7'2—dc20   95-22892
                                                  CIP

Acquisitions editor: Steve Chapman
Editorial team: Lori Flaherty, Executive Editor
              Andrew Yoder, Book Editor
Production team: Katherine G. Brown, Director
                Lisa M. Mellott, Coding
                Wanda S. Ditch, Desktop Operator
                Joann Woy, Indexer
Design team: Jaclyn J. Boone, Designer   9117201
             Katherine Lukaszewicz, Associate Designer   EL3

To Joyce, Geoff, and Amber

Contents

Introduction *xi*

1 Strategic Issues *1*

1.1 Why Code? *1*
1.2 Data Errors *1*
    1.2.1 Binary Symmetric Channel *1*
    1.2.2 Burst Channel *1*
    1.2.3 Binary Erasure Channel *2*
1.3 Software Notes *2*

2 Mathematical Tools for Coding *5*

2.1 Modulo Arithmetic *5*
    2.1.1 Modulo-2 Arithmetic *6*
2.2 Finite Fields *9*
    2.2.1 Groups *9*
    2.2.2 Fields *11*
2.3 Introduction to Polynomials *13*
2.4 Extension Fields *16*
2.5 Binary Extension Fields: Manual Construction *20*
    2.5.1 Constructing the Field *22*
    2.5.2 Representing the Field Elements *24*
2.6 Computer Methods for Galois Fields *26*
    2.6.1 Representing Field Elements *27*
2.7 More About Polynomials *30*
2.8 Cyclotomic Cosets *33*
2.9 Finding Minimal Polynomials *36*

3 Block and Cyclic Codes *41*

3.1 Introduction *41*
3.2 Linear Block Codes *44*

      3.2.1 Encoding for Linear Block Codes *44*
      3.2.2 Constructing the Generator Matrix *47*
      3.2.3 Parity-Check Matrix *47*
  3.3 Cyclic Codes *49*
  3.4 Manual Encoding Methods for Cyclic Codes *51*
  3.5 Modifications to Cyclic Codes *52*
      3.5.1 Extended Codes *53*
      3.5.2 Punctured Codes *54*
      3.5.3 Expurgated Codes *54*
      3.5.4 Augmented Codes *55*
      3.5.5 Shortened Codes *56*
      3.5.6 Lengthened Codes *56*
  3.6 Hamming Codes *56*
      3.6.1 Shortened Hamming Codes *58*

4 BCH and Reed-Solomon Codes *61*

  4.1 BCH Codes *61*
      4.1.1 Genesis of the Codes *62*
      4.1.2 Types of BCH Codes *63*
      4.1.3 Critical Features *63*
      4.1.4 Constructing the Generator Polynomial *68*
      4.1.5 Algorithmic Approach *71*
      4.1.6 Computer Approach *73*
  4.2 Nonbinary BCH Codes *79*
  4.3 Reed-Solomon Codes *82*

5 Encoders and Decoders *85*

  5.1 Division Method for Encoding Cyclic Codes *85*
      5.1.1 Hardware Implementation *85*
      5.1.2 Software Implementation *88*
  5.2 Standard Array *89*
      5.2.1 Hazards in Standard Array Decoding *92*
  5.3 Syndromes for BCH Codes *93*
  5.4 Peterson-Berlekamp Method *97*
  5.5 Error Location *102*

6 Convolutional Codes *105*

  6.1 Canonical Example *105*
      6.1.1 Convolutional Encoder Viewed as a Moore Machine *106*

6.1.2 Convolutional Encoder Viewed as a Mealy Machine *106*
6.1.3 Moore Machines vs. Mealy Machines *109*
6.1.4 Notation and Terminology *112*
6.2 Tree Representation of a Convolutional Encoder *116*
6.3 Trellis Representation of a Convolutional Encoder *119*
6.4 Distance Measures *123*

7 Viterbi Decoding *127*
7.1 Introduction to Viterbi Decoding *127*
7.2 Viterbi Decoding Failures *141*
7.2.1 Minimum Free Distance *146*
7.2.2 Weight Distribution *147*
7.2.3 Information Sequence Weight *147*
7.3 Viterbi Decoding with Soft Decisions *147*
7.4 Practical Issues *152*
7.4.1 Decoding Depth *153*

8 Sequential Decoding *155*
8.1 Stack Decoding Algorithm *155*
8.1.1 Fano Metric *161*
8.2 Software for Stack Decoding *168*
8.2.1 Implementing the Stack *171*
8.2.2 Received Symbol Buffering *177*
8.2.3 State Table *179*
8.3 Fano Decoding Algorithm *180*

Appendices

A Minimal Polynomials of Elements in GF(2^m) for Chapter 2 *191*

B Stack Tables for Example 8.1 *197*

C Stack Tables for Example 8.2 *207*

D Stack Tables for Example 8.3 *215*

E Software *227*

Index *245*

About the Author *251*

Disk Warranty *260*

Introduction

Error-correction coding has traditionally been one of the most mathematically challenging of the various specialized disciplines that come together to make modern data communication and data storage possible. A good bit of the difficulty no doubt stems from the use of a type of arithmetic that is different from the usual "standard" arithmetic that we all use every day. This book is an attempt to change all that. After Chapter 1 provides a brief introduction to the strategic issues involved in error correction coding, Chapter 2 demystifies those elements of abstract algebra and Galois field arithmetic that are essential for successful implementation of encoders and decoders. Some basic software modules for generating Galois fields and performing arithmetic within these fields are also developed in Chapter 2. Chapter 3 uses techniques from Chapter 2 to introduce the basics of block and cyclic codes. Chapter 4 covers the specific details of BCH codes and Reed-Solomon codes, which together account for perhaps 80% of all block codes used in practical applications. In Chapter 5, the details of encoders and decoders for block and cyclic codes are introduced and particularly efficient decoding techniques for BCH codes are covered in some detail. The major emphasis is on software implementations rather than hardware implementations, but digital hardware designers will be readily able to "port" the various software constructions into the corresponding hardware constructions. In addition to Block and Cyclic codes, there is another large class of codes called *convolutional codes*, which are also widely used in practical applications. Chapter 6 introduces convolutional codes and the notation and terminology commonly associated with them. Viterbi decoding of convolutional codes is covered in Chapter 7. This coverage includes the development of software modules for performing the Viterbi algorithm. Although it enjoys widespread use, Viterbi decoding does have some limitations. Because of these limitations, alternative decoding techniques, such as stack decoding or Fano decoding, must often be used. These alternative techniques, along with software for performing them, are covered in Chapter 8.

The subject of error correction coding is vast, and this book covers only a small part of it. However, the codes and techniques covered are the most widely used in practical applications. This, after all, accomplishes this book's original motive, which is to place powerful coding techniques as well as the knowledge needed to implement these techniques in the hands of individuals who need to use error-correction coding—even though these individuals are not coding experts.

Strategic Issues

1.1 Why Code?

Coding is performed for several different reasons. *Cryptographic* codes are used to protect information from unauthorized reception and to *authenticate* that a message was in fact sent by the party listed in the message as the originator. *Data compression* codes are used to reduce the amount of bandwidth needed to transmit data or to reduce the amount of memory needed to store a message. *Error control codes*, which are the subject of this book, are used to detect and/or correct errors that might be introduced into digital data when it is transmitted or stored.

1.2 Data Errors

There are several different mechanisms by which errors are introduced into a data signal. The strategy for coding the signal to protect it against errors is often tailored to the particular error mechanism known or assumed to be operating in application of interest.

1.2.1 Binary Symmetric Channel

As shown in Fig. 1-1, in a *binary symmetric channel (BSC)*, the probability of a 1 to 0 error and the probability of a 0 to 1 error are equal. This mechanism is most often assumed to be the dominant error-producing mechanism for errors produced by additive Gaussian noise. If the probability of error is independent from bit-to-bit, the BSC can be further characterized as being a *discrete memoryless channel (DMC)*.

1.2.2 Burst Channel

In many situations, the noise in a channel will vary over time, being weak during some situations and strong during other intervals. Errors will be more likely to occur in bursts that coincide with intervals of strong noise. Such a channel is referred to as a *burst channel*, or a *channel with memory*.

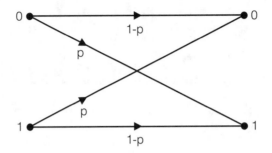

■ **1-1** *Transition probabilities in the binary symmetric channel.*

1.2.3 Binary Erasure Channel

Consider a baseband pulse transmission system in which a pulse of –1 volt is transmitted for a 0 and a pulse of +1 volt is transmitted for a 1. At the receiver, a pulse voltage less than –0.7 volt is interpreted as a 0. A pulse voltage greater than +0.7 is interpreted as a 1. The voltage range between –0.7 and +0.7 is considered as a "no-man's land" or *deadband*. Pulse voltages received in this range are deemed "too close to call" and are not interpreted as either 0 or 1. Such decisions not to decide is called *erasures*, and a channel that produces erasures is called a *binary erasure channel (BEC)*. The probability of a transmitted 0 being so corrupted that the received voltage exceeds 0.7 is so small that it is assumed to be impossible. Likewise for the probability of a transmitted 1 being so corrupted that the received voltage falls below –0.7. The probabilities of a 0-to-erasure failure or 1-to-erasure failure are not negligible, and special coding strategies are often used for correcting erasures. The BEC is depicted in Fig. 1-2, with erasures denoted as x.

1.3 Software Notes

The programs in this book were constructed as a set of *modules*. Modules have been described as "poor folks' objects"[1] and they are about as close as we can conveniently get to "true" objects without stepping up from **C** to **C++**. Modules support a subset of object-oriented programming that has been dubbed *object-based programming*. Modules don't support inheritance, but they do support abstraction and encapsulation. Modules are implemented in **C** as separate source files. This means that the programs in this book

1 McConnell: *Code Complete*, Microsoft Press, 1993.

2

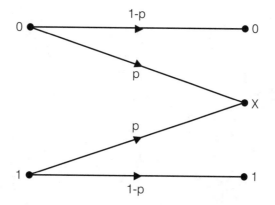

■ **1-2** *Transition probabilities in the binary erasure channel.*

are organized into modules that make sense in the context of what we're trying to accomplish, but which do not necessarily match up with the way chapter contents are organized.

Each file contains both data and functions which can be declared static to make them visible only within the file. *Module variables* are static variables, which are global **within the file**. Not every function in a module makes use of every module variable, so as each function is presented in the text, in the way that it interacts with module variables is also covered. The software accompanying this book is described in Appendix D.

3

Mathematical Tools for Coding

2

Many of the calculations used to perform encoding, decoding, and performance analysis for the codes in this book make use of peculiar arithmetic and other mathematical techniques that are a bit out of the ordinary. These techniques, which are essential for the work in subsequent chapters, include modulo-p arithmetic, extension fields, and polynomial arithmetic. These topics sound a lot more difficult than they really are, and this chapter attempts to introduce and explain this material in a way that is relatively painless. This goal is consistent with the hands-on "cookbook" approach chosen for this book. However, presenting these techniques as isolated, unconnected rote procedures will prove somewhat unsatisfying to many readers and might actually make it more difficult for these readers to embrace these new techniques and begin using them. Therefore, this chapter is a compromise between extremes. The individual sections present the necessary foundation material embedded in a broader mathematical fabric for interested readers. However, wherever it is appropriate, the sections begin with a box entitled "The Basic Idea." The contents of this box briefly explains how the subsequent material in the section fits into the overall study of coding with an emphasis on why it is important for practical coding applications. Most sections end with a box entitled "The Bottom Line." This box summarizes the information that the reader needs to become familiar with and carry forward into subsequent sections and chapters.

2.1 Modulo Arithmetic

The Basic Idea

Many error detection and correction (EDAC) codes are based on peculiar types of arithmetic that make use of mathematical structures called *fields*, which is covered in Section 2.2.

Two specific types of fields—*prime fields* and *extension fields*—both make extensive use of modulo arithmetic. Therefore, this section reviews modulo arithmetic as preparation for exploring fields.

Modulo arithmetic can be viewed as the result of "wrapping" the integers around a circle (Fig. 2-1) for the specific case of modulo-5 arithmetic. This wrapping causes 5 to be "the same as" or *congruent* to 0. This forces 6 to be congruent to 1, 7 to be congruent to 2, 10 to be congruent to 0, and so on. The congruences need not be limited to just positive integers. As shown in the figure, –1 is congruent to 4, –3 is congruent to 2, etc. The usual mathematical notation for congruence is a triple-bar equals sign, "≡." Thus "1 ≡ 16" is read as "1 is congruent to 16."

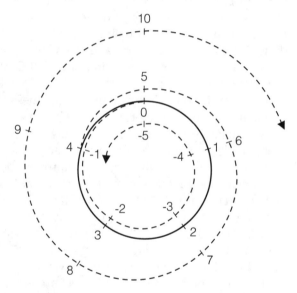

■ **2-1** *The modulo-5 number system can be viewed as the set of integers "wrapped" around a circle to establish congruences between 0 and 5, 1 and 6, –1 and 4, etc.*

2.1.1 Modulo-2 Arithmetic

The vast majority of the computations used to perform encoding and decoding involve modulo-2 arithmetic. This is due primarily to the close relationship between modulo-2 arithmetic and the *Boolean algebras* commonly used by designers of digital logic circuits.

Boolean Algebras

Encoders and decoders (not to mention computers) are constructed from logic circuits of various types. The logical behavior of these circuits can be modeled and analyzed using a particular type of algebra called *Boolean algebra* (which is named in honor of its inventor George Boole). Strictly speaking, Boolean algebra includes several different algebras, each of which uses different sets of logical operations, such as {AND, OR, INVERT} or {NAND, NOR, INVERT}. In the particular Boolean algebra of interest here, there are two *elements* or *values* as well as two logical *operations*. The values are TRUE or FALSE, and the operations are AND, and XOR (exclusive or). The results of these operations are listed in Tables 2-1 and 2-2.

■ **Table 2-1 The logical AND operation of Boolean algebra.**

AND	F	T
F	F	F
T	F	T

■ **Table 2-2 The logical XOR operation of Boolean algebra.**

XOR	F	T
F	F	T
T	T	F

Modulo-2 Arithmetic

A mathematical analog or *isomorphism* to the logical-valued Boolean algebra of Tables 2-1 and 2-2 uses values of 0 and 1, and operations of modulo-2 multiplication and addition, as shown in Tables 2-3 and 2-4. The value 0 corresponds to FALSE, and the value 1 corresponds to TRUE. Modulo-2 multiplication corresponds to the logical AND operation. The result of (a AND b) is true only if a and b are both TRUE.

Similarly, the result of $a \times b$ is 1 only if a and b both equal 1. Modulo-2 addition corresponds to the logical XOR operation. The result of (a XOR b) is 1 only if either a or b (but not both) is TRUE. The result of $a + b$ is 1 only if either a or b (but not both) is 1.

■ Table 2-3
Binary multiplication.

×	0	1
0	0	0
1	0	1

■ Table 2-4
Modulo-2 addition.

+	0	1
0	0	1
1	1	0

Programming Considerations Modulo arithmetic can be implemented in **C** for arbitrary moduli using the *remainder* operator. The expression x%y returns the remainder produced by dividing x by y. Assuming that x and y are both "legal" values modulo-q (*i.e.* $0 \leq x < q$ and $0 \leq x < q$) then modulo-q addition and multiplication can be easily implemented, as shown in the following code fragment.

```
int x, y, q, sum, product
/***********************************/
/*  Note: if q==0, a divide-by-zero  */
/*  error will occur.                */

/*  modulo-q addition  */
sum = (x+y) %q;

/*  modulo-q multiplication  */
product = (x*y) %q;

/**  end of fragment  **/
```

Modulo-2 arithmetic occurs so frequently that it might lead to greater program efficiency if we use Boolean operations in place of the remainder operation for modulo-2 arithmetic. As shown in the following code fragment, there are a number of possibilities.

```
int x, y, q, sum1, sum2, sum3, prod1, prod2

/*  modulo-2 addition  */

sum1 = x ^ y;      /*  uses bitwise XOR  */
sum2 = x != y;     /*  uses NOT EQUALS  */
sum3 = !(x && y);  /*  uses logical NOT with logical AND
                       to implement logical NAND  */
```

```
/* modulo-2 multiplication */
prod1 = x & y      /* uses bitwise AND */
prod2 = x && y     /* uses logical AND */

/** end of fragment **/
```

2.2 Finite Fields

> ### The Basic Idea
>
> As so eloquently stated by R. J. McEliece,[1] "A field is a place
> where you can add, subtract, multiply and divide." Ordinary
> arithmetic takes place in an *infinite field*, so named because
> it has an infinite number of elements. Coding applications
> make use of arithmetic in a *finite field*. This section exam-
> ines just what it takes for a collection of elements and opera-
> tions to constitute a field. The related concept of a *group* is
> presented first because a field is most easily defined in terms
> of groups.

2.2.1 Groups

Modulo-2 addition is an example of a mathematical structure
called a *group*, and we can use modulo-2 addition to illustrate a
number of important properties that are shared by all groups:

1. The set of integers {1,0} is said to be *closed* under the
   operation of modulo-2 addition, because the modulo-2
   addition of any two elements in the set produces a result that
   is also in the set.

2. The modulo-2 addition operation is *associative*, meaning that
$$(a + b) + c = a + (b + c)$$
   for all a, b, and c which are elements of {0,1}.

3. The element 0 is called the *identity element* and has the
   property that
$$a + 0 = 0 + a = a$$
   for all a, which are elements of {0,1}.

4. For each element a in {0,1}, there is an element $-a$ that when
   added to a produces a result of 0:
$$0 + 0 = 0 \qquad 1 + 1 = 0$$
   The element $-a$ is called the *additive inverse of* a. (In
   modulo-2 addition, each element is its own inverse, but in
   general, this will not be true of other groups.)

1 R. J. McEliece: *Encyclopedia of Mathematics and Its Applications, Vol. 3: The
Theory of Information and Coding,* Addison-Wesley, Reading MA, 1977.

9

The group formed by modulo-2 addition is further identified as a *commutative group* or *abelian group* because $a + b = b + a$ for all a and b, which are elements of {0,1}.

It turns out that modulo-q addition forms a commutative group for any value of $q = 2,3,4,...$

Example 2.1 Show that modulo-3 addition forms a commutative group.

Solution Modulo-3 addition results are summarized in Table 2-5.

■ Table 2-5
Modulo-3 addition.

+	0	1	2
0	0	1	2
1	1	2	0
2	2	0	1

Each of the five prerequisites for a commutative group are satisfied:

☐ Inspection of Table 2-5 reveals that modulo-3 addition is closed. All results are elements of the set {0,1,2}.

☐ Modulo-3 addition is associative.

☐ The identity element for modulo-3 addition is 0.

☐ Inspection of Table 2-5 reveals that each element has an additive inverse. Specifically,

$$-0 = 0 \qquad -1 = 2 \qquad -2 = 1$$

\end{array}

☐ Table 2-5 is symmetric about its main diagonal; therefore, the operation is commutative.

Modulo-3 **multiplication** also forms a commutative group over the nonzero elements {1, 2}. The results for modulo-3 multiplication are summarized in Table 2-6.

■ Table 2-6
Modulo-3 multiplication.

×	0	1	2
0	0	0	0
1	0	1	2
2	0	2	1

The five prerequisites (some of them reworded slightly for multiplication) are satisfied:

- [ ] Inspection of the table reveals that modulo-3 multiplication is closed over $\{1, 2\}$.
- [ ] Modulo-3 multiplication is associative.
- [ ] The identity element for multiplication (or *multiplicative identity*) is 1.
- [ ] Inspection of Table 2-6 reveals that each element in $\{1, 2\}$ has a multiplicative inverse. Specifically,

$$1 \times 1 = 1 \qquad 2 \times 2 = 1$$

- [ ] In algebraic form, the multiplicative inverse of a is denoted as a^{-1}.

  (Note: In general, each element will not be its own multiplicative inverse as it is here. For example, in modulo-5 multiplication, the multiplicative inverse of 2 is 3.)

It turns out that the set of **nonzero** elements $\{1, 2,..., p - 1\}$ modulo-p with multiplication forms a group over for all prime values of p. However, for composite (*i.e.*, nonprime) values of q, modulo-q multiplication does not form a group over the set $\{1, 2,..., q - 1\}$.

Example 2-2 Show that modulo-4 multiplication does not form a group over the elements $\{1, 2, 3\}$.

Solution Modulo-4 multiplication results are summarized in Table 2-7. The set of nonzero values $\{1, 2, 3\}$ is not closed under modulo-4 multiplication because examination of the table reveals that $2 \times 2 = 0$. Furthermore, there is no multiplicative inverse for the element 2.

■ **Table 2-7**
Modulo-4 multiplication.

×	0	1	2	3
0	0	0	0	0
1	0	1	2	3
2	0	2	0	2
3	0	3	2	1

2.2.2 Fields

A set F, in conjunction with two operations (one "addition-like" called addition and one "multiplication-like" called multiplication) defined over the elements of F forms a mathematical structure called a *field* if and only if the following conditions are satisfied:

1. The elements of F form a **commutative group** under the operation of addition.

2. The nonzero elements of F (that is all elements except the additive inverse) form a **commutative group** under the operation of multiplication.

3. The multiplication operation **distributes** over the addition operation.

The number of elements in a field can be either finite or infinite. The set R of real numbers, in conjunction with ordinary addition and multiplication, is an example of an infinite field. In coding theory, we will be concerned exclusively with **finite** fields.

Definition 2.1 The number of elements in a field is called the *order* of the field.

Finite fields are also called *Galois fields* in honor of Évariste Galois who outlined the connection between groups and polynomial equations in a letter to his friend Auguste Chevalier written on the eve of a duel on May 30, 1832 in which Galois was fatally wounded.[2] A Galois field of order q is denoted as $GF(q)$. The field specified by Tables 2-3 and 2-4 is thus denoted as $GF(2)$. This field has 2 elements and makes use of modulo-2 addition and modulo-2 multiplication. It is possible to extend this idea to a field with p elements using modulo-p arithmetic, provided that p is prime. Such a field is called a *prime field*. As you will see in Section 2.4, it is also possible to construct fields $GF(q)$, where $q = p^m$, $m = 1, 2,...$

The Bottom Line

☐ Groups are a set of elements with an operation defined for these elements such that:
  ~ The operation is **closed** over the set of elements.
  ~ The operation is **associative.**
  ~ The set contains an **identity element.**
  ~ For each element in the set, there is an **inverse** also contained in the set.

☐ If a group's operator is commutative, the group is called a **commutative group** or **abelian group.**

☐ Modulo-q addition forms a commutative group for all $q \geq 2$.

2 I. Stewart: *Galois Theory*, 2nd ed., Chapman & Hall, London, 1989.

- ☐ Modulo-p multiplication forms a commutative group over the nonzero elements $\{1, 2,..., p-1\}$ for all **prime** values of p.
- ☐ Modulo-q multiplication does not form a group for nonprime q.
- ☐ Fields are a set of element F with two operations— "addition" and "multiplication"—defined for these elements such that:
  - ~ The elements of F form a commutative group under addition.
  - ~ The nonzero elements of F form a commutative group under multiplication.
  - ~ Multiplication distributes over addition.
- ☐ Finite fields are also called **Galois fields** and are therefore denoted as GF(p).

2.3 Introduction to Polynomials

The Basic Idea

This section presents a review of polynomial arithmetic for polynomials whose coefficients are drawn from prime fields and which require only modulo arithmetic for operating on these coefficients. This background is needed before we can explore the construction of extension fields in Section 2.4. Ultimately, we will be concerned with polynomials whose coefficients are drawn from extension fields, and which therefore require extension-field arithmetic for operating on these coefficients. This second type of polynomial is covered in Section 2.7.

13

Much of coding theory (including the construction of extension fields to be covered shortly) is based on *polynomial arithmetic*. A *polynomial in x* is simply a weighted sum of powers of x. Some examples of polynomials in x are

$$f_1(x) = 5x^2 + 3x + 1$$
$$f_2(x) = x^4 + x^3 + 1$$
$$f_3(x) = x^3 - 2$$

The degree of the polynomial is the highest power of x appearing in the polynomial. The degree of $f_1(x)$ is 2; the degree of $f_2(x)$ is 4; and the degree of $f_3(x)$ is 3.

In coding theory, we will often be interested in polynomials in which the coefficient of each term is either 0 or 1. In the context of coding theory, such a polynomial is called a *polynomial over GF(2)*. Of the three example polynomials shown, $f_2(x)$ is the only polynomial over GF(2). When performing arithmetic on polynomials over GF(2), powers of x can be multiplied in the usual way; that is,

$$x^2 \bullet x^3 = x^5$$
$$x \bullet x^7 = x^8$$
$$x \bullet x = x^2$$

However, the addition of coefficients is performed using modulo-2 addition. For example,

$$x^2 + x^2 = 1 \bullet x^2 + 1 \bullet x^2 = (1 + 1) \bullet x^2 = 0 \bullet x^2 = 0,$$

$$(x + 1) + (x^2 + x) = 1 \bullet x + 1 + 1 \bullet x^2 + 1 \bullet x$$
$$= 1 \bullet x^2 + (1 + 1) \bullet x + 1$$
$$= 1 \bullet x^2 + 0 \bullet x + 1 = x^2 + 1$$

and

$$x^5 + x^5 + x^5 = (1 + 1 + 1) \bullet x^5 = 1 \bullet x^5 = x^5$$

Because each element in GF(2) is its own inverse, subtraction of polynomials over GF(2) is the same as addition:

$$x^3 + x + 1 = x^3 + x + 1$$
$$-(x^3 + x^2 + 1) = \frac{+(x^3 + x^2 + 1)}{x^2 + x + 1}$$

However, this is not true for polynomials over GF(p) $p > 2$. Consider the subtraction of two polynomials over GF(3):

$$x^3 + x + 1 = x^3 + x + 1$$
$$-(2x^3 + x^2 + 2) = \frac{+(x^3 + 2x^2 + 1)}{2x^3 + 2x^2 + x + 2}$$

[Note that in GF(3), $-2 = 1$, $-1 = 2$.]

A general polynomial of degree N can be written as

$$f(x) = b_N x^N + b_{N-1} x^{N-1} + ... + b_2 x^2 + b_1 x + b_0 \qquad (2.1)$$

where some of the b_n might be zero for $n < N$. (If $b_N = 0$, the degree of the polynomial is not N.) A more compact notation is

$$f(x) = \sum_{n=0}^{N} b_n x^n \qquad (2.2)$$

Programming Considerations How can a polynomial such as Eqs. 2.1 or 2.2 be represented in software? We can store the coefficients of $f(x)$ in an array, for example coef[], with the coefficient for x^n in location n. When the polynomial is represented this way, it is easy to multiply or divide the polynomial by a power of x. To multiply $f(x)$ by x^k, for $k \geq 0$, simply shift the coefficients upward by k locations in the array, as shown in the following code fragment:

```
/**  fragment to multiply polynomial by x**k  **/

/*  shift original N coefficients  */

for( n=N; n>=0; n-- )

    {
    coef[n+k]  =  coef[n];
    }

/*  store zeros for the k lowest-order new coefficients  */

for( n=k-1; n>=0; n-- )
    {
    coef[n]  =  0;
    }

/*  increase degree of polynomial by k  */

N  += k;

/**  end of fragment  **/
```

To divide $f(x)$ by x^k for $k \geq 0$, shift the coefficients downward by k locations in the array as shown in the following program fragment. The k lowest-order terms in the original polynomial will become the remainder of the division operation.

```
/**  fragment to divide polynomial by x**k  **/

/*  save remainder  */

for( n=0; n<k; n++ )
    {
    remainder[n]  =  coef[n];
    }

/*  downshift the higher-order coefficients by k places  */

for( n=0; n<=(N-k); n++ )
    {
    coef[n]  =  coef[n+k];
    }

/*  store zeros in locations originally occupied by k
higher-order terms  */
```

```
for( n=(N-k+1); n<=N; n++)
    {
    coef[n] = 0;
    }

/** end of fragment **/
```

2.4 Extension Fields

The Basic Idea

Finite fields based solely on modulo arithmetic are restricted to have prime numbers of elements. This section develops a specific *extension field*, which has $2^3 = 8$ elements. The principles used in this section are generalized in Section 2.5 to include extension fields having 2^m elements where m is any positive integer. Extension fields, especially those having 2^m elements, are widely used in coding applications.

As demonstrated in Section 2.1, it is a simple matter to construct a field GF(p) for any prime p by using modulo-p addition and multiplication. In working with various codes, it turns out to be very useful if all of the necessary manipulations can be performed within a field having 2^m elements, with each element being an m-bit binary value. But so far, we have seen only fields in which the number of elements is a prime number; and of course, 2^m is not prime for $m > 1$. What happens if we try to construct a field with 2^m elements using modulo-2^m arithmetic?

Example 2.3 Consider the specific case of $2^m = 2^3 = 8$. The operations of modulo-8 addition and modulo-8 multiplication are defined in Tables 2-8 and 2-9. Examination of Table 2-8 reveals that modulo-8 addition does form a commutative group (see Section 2.2.2). However, examination of Table 2-9 reveals that there are no multiplicative inverses for the even numbers 2, 4, and 6. This means that the integers 1, 2, 3, 4, 5, 6, and 7 do not form a group under modulo-8 multiplication; therefore, modulo-8 arithmetic cannot be used to construct a field GF(2^m).

Although it is not possible to construct a Galois field GF(2^m) using modulo-2^m multiplication, it is possible to construct such a field using *polynomial arithmetic*.

■ Table 2-8 Modulo-8 addition.

+	0	1	2	3	4	5	6	7
0	0	1	2	3	4	5	6	7
1	1	2	3	4	5	6	7	0
2	2	3	4	5	6	7	0	1
3	3	4	5	6	7	0	1	2
4	4	5	6	7	0	1	2	3
5	5	6	7	0	1	2	3	4
6	6	7	0	1	2	3	4	5
7	7	0	1	2	3	4	5	6

■ Table 2-9 Modulo-8 multiplication.

×	0	1	2	3	4	5	6	7
0	0	0	0	0	0	0	0	0
1	0	1	2	3	4	5	6	7
2	0	2	4	6	0	2	4	6
3	0	3	6	1	0	3	6	1
4	0	4	0	4	0	4	0	4
5	0	5	2	7	4	1	6	3
6	0	6	4	2	0	6	4	2
7	0	7	6	5	4	3	2	1

Another Way to Create Fields

So far we have considered fields in which the elements are integers and the operations are modulo-p addition and modulo-p multiplication. Except for the "wrap-around" that occurs between $p - 1$ and 0, these operations are closely related to the usual arithmetic operations of addition and multiplication. In order to construct fields having a composite (nonprime) number of elements, we must resort to operations that are not so closely related to the operations of ordinary arithmetic. Despite their unfamiliarity, these operations are not difficult; and virtually all of coding theory is based on fields that are built upon a few simple rules.

Our principal applications for error-correction coding will involve digital values in binary form. Therefore, we will focus on finding a way to construct finite fields having 2^m elements with each element being an m-bit binary value.

Consider a commutative group based on elements that are 3-bit binary values and the operation of bit-by-bit modulo-2 addition (or equivalently, bit-by-bit logical exclusive OR). The results of this operation are listed in Table 2-10. All of the requirements for a group are satisfied:

☐ The operation is associative.

☐ The identity element is 000.

☐ Each element is its own inverse.

■ Table 2-10 Bit-by-bit modulo-2 addition of 3-bit values.

	000	001	010	011	100	101	110	111
000	000	001	010	011	100	101	110	111
001	001	000	011	010	101	100	111	110
010	010	011	000	001	110	111	100	101
011	011	010	001	000	111	110	101	100
100	100	101	110	111	000	001	010	011
101	101	100	111	110	001	000	011	010
110	110	111	100	101	010	011	000	001
111	111	110	101	100	011	010	001	000

Assume that this is the additive group for our desired field. We must now find a "multiplication-like" operation that forms a commutative group over the nonzero elements 001, 010, 011, 100, 101, 110, and 111. By "multiplication-like," we mean that whatever this operation turns out to be, it must distribute over the bit-by-bit modulo-2 addition defined by Table 2-10. An operation that meets all of the requirements is defined by Table 2-11. What are the rules used to define this table? None of the logical operations—AND, OR, NAND, NOR, etc.—fits the bill. To find suitable rules, we need to turn to polynomial arithmetic. Let each value in Table 2-11 represent a polynomial in α. Each bit within a value represents the coefficient of a particular power of α. Specifically,

$$111 \text{ represents } \alpha^2 + \alpha + 1$$
$$110 \text{ represents } \alpha^2 + \alpha$$
$$101 \text{ represents } \alpha^2 + 1$$
$$100 \text{ represents } \alpha^2$$
$$011 \text{ represents } \alpha + 1$$
$$010 \text{ represents } \alpha$$
$$001 \text{ represents } 1$$
$$000 \text{ represents } 0$$

	001	010	011	100	101	110	111
001	001	010	011	100	101	110	111
010	010	100	110	011	001	111	101
011	011	110	101	111	100	001	010
100	100	011	111	110	010	101	001
101	101	001	100	010	111	011	110
110	110	111	001	101	011	010	100
111	111	101	010	001	110	100	011

We can try defining a multiplicative operation over this set of polynomials using the ordinary rules of algebraic multiplication. For example

$$\alpha \bullet (\alpha + 1) = \alpha^2 + \alpha \Leftrightarrow (010) \bullet (011) = (110)$$

However, multiplication defined in this way is not closed over the set of 3-bit elements. For example

$$\alpha \bullet (\alpha^2 + 1) = \alpha^3 + \alpha^2$$

and we have no way to represent α^3 in our 3-bit scheme. In much the same way that we define $p \equiv 0$ to force closure in modulo-p arithmetic, we can define $\alpha^3 \equiv \alpha + 1$. This relationship can be used to express α^3, α^4, α^5,... in terms of 1, α, and α^2. For example, we can write

$$\alpha^3 = \alpha + 1 \text{ (by definition)}$$
$$\alpha^4 = \alpha \bullet \alpha^3 = \alpha(\alpha + 1) = \alpha^2 + \alpha$$
$$\alpha^5 = \alpha \bullet \alpha^4 = \alpha(\alpha^2 + \alpha) = \alpha^3 + \alpha^2$$
$$= (\alpha + 1) + \alpha^2 = \alpha^2 + \alpha + 1$$
$$\alpha^6 = \alpha \bullet \alpha^5 = \alpha(\alpha^2 + \alpha + 1) = \alpha^3 + \alpha^2 + \alpha$$
$$= (\alpha + 1) + \alpha^2 + \alpha = \alpha^2 + 1$$

Table 2-12 lists elements α^0 through α^{10} expressed in terms of 1, α, and α^2. Looking at the table, we see that $\alpha^7 = \alpha^0$, $\alpha^8 = \alpha^1$, $\alpha^9 = \alpha^2$, etc. Similarly to the way that defining $5 \equiv 0$ created a closed finite subset of the integers used to construct GF(5), defining $\alpha^3 = \alpha + 1$ has created a closed finite subset of the powers of α. Therefore, we can conclude that the operation at work in Table 2-11 is the special form of polynomial multiplication, based upon the relationship $\alpha^3 = \alpha + 1$. Tables 2-10 and 2-11 comprise exactly the sort of thing we said we needed—a field containing 2^m elements where each element has an m-bit binary value. Such fields are called *extension fields* of GF(2), and the field GF(2) is called the *ground field* or *prime field* of GF(2^m).

19

■ Table 2-12 Three different
representations for GF(2^3).

Power form	Polynomial form	3-tuple form
0	0	000
1	1	001
α	α	010
α^2	α^2	100
α^3	$\alpha + 1$	011
α^4	$\alpha^2 + \alpha$	110
α^5	$\alpha^2 + \alpha + 1$	111
α^6	$\alpha^2 + \quad 1$	101
α^7	1	001
α^8	α	010
α^9	α^2	100
α^{10}	$\alpha + 1$	011

The Bottom Line

Using only modulo arithmetic, it is not possible to construct fields having nonprime numbers of elements. However, this section has shown that it is possible to use a special form of polynomial arithmetic to build the extension field GF(2^3), which has 2^3 elements. In Section 2.5, we tackle the problem of how to extend our results for GF(2^3) to the general case of GF(2^m).

2.5 Binary Extension Fields: Manual Construction

The Basic Idea

Making use of *primitive polynomials*, it is possible to define an extension field GF(2^m) for any integer $m \geq 2$. Such fields will be used extensively throughout the rest of this book. This might be one of the most difficult sections in the book, but it is also one of the most important!

If q is not prime, the set of elements $\{1,2,...,q-1\}$ does not form a group under modulo-q multiplication; therefore, modulo-q multiplication cannot be used to define a field GF(q). However, in Section 2.4, we found that it is possible to define a field GF(2^3). In fact, it is possible for **any** prime p to define fields having p^m elements by using polynomial arithmetic in place of modulo arithmetic. However, in this book, we are primarily interested in extension fields of the form $GF(2^m)$ having 2^m elements. A particular type of polynomial called a *primitive polynomial*, plays a role in the construction of extension fields that is similar to the role played by the modulus in construction of prime fields.

Definition 2.2 Let $p(x)$ be a polynomial of degree m over $GF(2)$. If $p(x)$ is not divisible by any polynomial over $GF(2)$ of degree one through $m - 1$, then $p(x)$ is *irreducible* over $GF(2)$.

Reality Check

A polynomial of degree m over $GF(2)$ is a polynomial of degree m in which each coefficient is an element of $GF(2)$. The polynomial $p(x) = x^3 + x + 1$ is *irreducible* over $GF(2)$ because $p(x)$ is not divisible by any of the polynomials of degree 2 over $GF(2)$—$x^2, x^2 + 1, x^2 + x, x^2 + x + 1$—nor by any of the polynomials of degree 1 over $GF(2)$—$x, x + 1$. On the other hand, the polynomial $x^3 + x$ is not irreducible because it is factorable into $(x^2 + x)(x + 1)$.

Result 2.1 Any irreducible polynomial over $GF(2)$ of degree m divides $x^{2m} - 1 + 1$.

Reality Check

We said that $x^3 + x + 1$ is an irreducible polynomial of degree 3 over $GF(2)$, so it should divide $x^{2^3 - 1} + 1 = x^7 + 1$. And if we attempt the division, we find that

$$\frac{x^7 + 1}{x^3 + x + 1} = x^4 + x^2 + x + 1$$

Definition 2.3 We know from Result 2.1 that if $p(x)$ is an irreducible polynomial over $GF(2)$ of degree m, then $p(x)$ divides $x^n + 1$ for $n = 2^m - 1$. If this value of n is the smallest positive integer for which $p(x)$ divides $x^n + 1$, then $p(x)$ is a *primitive polynomial of degree m* over $GF(2)$.

Reality Check

The polynomial $p_1(x) = x^3 + x + 1$ is a *primitive polynomial of degree* 3 over $GF(2)$, because $p_1(x)$ divides $x^7 + 1$ and it does not divide $x^6 + 1$, $x^5 + 1$, $x^4 + 1$, or $x^3 + 1$. On the other hand, the polynomial $p_2(x) = x^4 + x^3 + x^2 + x + 1$ is an irreducible polynomial of degree 4 over $GF(2)$ because $p_2(x)$ divides $x^{2^4 - 1} + 1 = x^{15} + 1$. However, $p_2(x)$ is not a primitive polynomial because it also divides $x^{10} + 1$ and $x^5 + 1$.

Binary Extension Fields: Manual Construction

Primitive polynomials over GF(2) for $2 \leq m \leq 30$ are listed in Appendix C.

2.5.1 Constructing the Field

For the case of binary coefficients, there are 2^m different polynomials of degree $\geq m - 1$. The field GF(2^m) will have 2^m elements, each of which will be one of the possible 2^m polynomials of degree $m - 1$. In order to preserve the properties of addition and multiplication that are needed in order to have a field, each of the 2^m polynomials must be assigned to the elements of GF(2^m) in a particular order. Recipe 2.1 provides a means for making this assignment correctly.

Recipe 2.1 Constructing the binary extension field GF(2^m). The complete extension field will consist of 2^m elements whose power forms are 0, 1, α, α^2, α^3,...,α^{2m-2}. The purpose of this recipe is to generate the polynomial forms for each of these elements.

1. Each of the elements 0, 1, α, α^2,...,α^{m-1} has a polynomial form that is exactly the same as its power form.

2. Select a primitive polynomial of degree m over GF(2). (In other words, select a primitive polynomial of degree m having coefficient values of 0 or 1.) Let $p(x)$ denote the primitive polynomial selected:

$$p(x) = x^m + f_{m-1}x^{m-1} + f_{m-2}x^{m-2} + \ldots + f_2 x^2 + f_1 x + f_0$$

3. Substitute α for x, set $p(\alpha) = 0$ and solve for α^m

$$\alpha^m = f_{m-1}\alpha^{m-1} + f_{m-2}\alpha^{m-2} + \ldots + f_2 \alpha^2 + f_1 \alpha + f_0 \qquad (2.3)$$

4. For $k = m, m + 1, m + 2, \ldots, 2^m - 3$, we can obtain a polynomial expression for α^{k+1} by multiplying the polynomial for α^k by α:

$$\alpha^{k+1} = \alpha \cdot \alpha^k$$

Sometimes the result will contain a term for α^m and sometimes the degree of the highest-degree term will be $m - 1$ or less. When the result does not contain a term for α^m, the result is the correct polynomial form for the α^{k+1} element. When the result does contain a term for α^m, substitute the polynomial form for α^m (Eq. 2.3) and simplify to obtain the polynomial form for α^{k+1}.

Recipe 2.1 can be used directly to construct an extension field GF(2^m) for any m, but it can also be used as the theoretical basis for a more easily mechanized procedure that will be presented in

Section 2.6. We can see now that the relationship $\alpha^3 = \alpha + 1$ that we used in Section 2.4 is derived from $p(x) = x^3 + x + 1$, which is a primitive polynomial of degree 3 over GF(2).

Example 2.4 Construct an extension field GF(2^4).

Solution

1. The first five elements are 0, 1, α, α^2, and α^3.
2. A primitive polynomial of degree 4 over GF(2) is p(x) = $x^4 + x + 1$.
3. $p(\alpha) = \alpha^4 + \alpha + 1 = 0 \Rightarrow \alpha^4 = \alpha + 1$

   A. $\alpha^5 = \alpha \cdot \alpha^4 = \alpha(\alpha + 1) = \alpha^2 + \alpha$

   B. $\alpha^6 = \alpha \cdot \alpha^5 = \alpha(\alpha^2 + \alpha) = \alpha^3 + \alpha^2$

   C. $\alpha^7 = \alpha \cdot \alpha^6 = \alpha(\alpha^3 + \alpha^2) = \alpha^4 + \alpha^3$
   substituting $\alpha^4 = \alpha + 1$ yields
   $\alpha^7 = \alpha^3 + \alpha + 1$

   D. $\alpha^8 = \alpha \cdot \alpha^7 = \alpha(\alpha^3 + \alpha + 1) = \alpha^4 + \alpha^2 + \alpha$
   substituting $\alpha^4 = \alpha + 1$ yields
   $\alpha^8 = \alpha^2 + 1$

   E. $\alpha^9 = \alpha \cdot \alpha^8 = \alpha(\alpha^2 + 1) = \alpha^3 + \alpha$

   F. $\alpha^{10} = \alpha \cdot \alpha^9 = \alpha(\alpha^3 + \alpha) = \alpha^4 + \alpha^2$
   substituting $\alpha^4 = \alpha + 1$ yields
   $\alpha^{10} = \alpha^2 + \alpha + 1$

   G. $\alpha^{11} = \alpha \cdot \alpha^{10} = \alpha(\alpha^2 + \alpha + 1) = \alpha^3 + \alpha^2 + \alpha$

   H. $\alpha^{12} = \alpha \cdot \alpha^{11} = \alpha(\alpha^3 + \alpha^2 + \alpha) = \alpha^4 + \alpha^3 + \alpha^2$
   substituting $\alpha^4 = \alpha + 1$ yields
   $\alpha^{12} = \alpha^3 + \alpha^2 + \alpha + 1$

   I. $\alpha^{13} = \alpha \cdot \alpha^{12} = \alpha(\alpha^3 + \alpha^2 + \alpha + 1) = \alpha^4 + \alpha^3 + \alpha^2 + \alpha$
   substituting $\alpha^4 = \alpha + 1$ yields
   $\alpha^{13} = \alpha^3 + \alpha^2 + 1$

   J. $\alpha^{14} = \alpha \cdot \alpha^{13} = \alpha(\alpha^3 + \alpha^2 + 1) = \alpha^4 + \alpha^3 + \alpha$
   substituting $\alpha^4 = \alpha + 1$ yields
   $\alpha^{14} = \alpha^3 + 1$

These results are summarized in Table 2-13.

■ Table 2-13 Three different representations for GF(2^4).

Power form	Polynomial form	4-tuple form
0	0	0000
1	1	0001
α	α	0010
α^2	α^2	0100
α^3	α^3	1000
α^4	$\alpha + 1$	0011
α^5	$\alpha^2 + \alpha$	0110
α^6	$\alpha^3 + \alpha^2$	1100
α^7	$\alpha^3 + \alpha + 1$	1011
α^8	$\alpha^2 + 1$	0101
α^9	$\alpha^3 + \alpha$	1010
α^{10}	$\alpha^2 + \alpha + 1$	0111
α^{11}	$\alpha^3 + \alpha^2 + \alpha$	1110
α^{12}	$\alpha^3 + \alpha^2 + \alpha + 1$	1111
α^{13}	$\alpha^3 + \alpha^2 + 1$	1101
α^{14}	$\alpha^3 + 1$	1001

2.5.2 Representing the Field Elements

Each element of the extension field GF(p^m) can be represented by an m-digit base-p value. For the specific case of GF(2^m), this means that each element is represented by an m-bit binary value. Each bit within a value corresponds to a coefficient in a polynomial of degree $m - 1$. Obviously, for such a representation to be unambiguous, the bits must correspond to the coefficients of the polynomials when the terms of the polynomials are in some specified order. Four obvious schemes can be used for this representation:

1. The polynomial is arranged from lowest-degree term to highest-degree term with the leftmost bit assigned to the α^0 term. For $m = 4$, this approach yields:

$$1 + \alpha + \alpha^2 + \alpha^3 \leftrightarrow 1111$$
$$1 + \alpha \leftrightarrow 1100$$
$$1 \leftrightarrow 1000$$
$$\alpha + \alpha^3 \leftrightarrow 0101$$

This approach is used by Rhee,[3] Lin and Costello,[4] and by MacWilliams and Sloane.[5]

3 M. Y. Rhee: *Error Correcting Coding Theory*, McGraw-Hill, New York, 1989.

4 S. Lin and D. J. Costello: *Error Control Coding: Fundamentals and Applications*, Prentice-Hall, Englewood Cliffs, NJ, 1983.

5 F. J. MacWilliams and J. J. A. Sloane: *The Theory of Error-Correcting Codes*, North Holland, Amsterdam, 1977.

2. The polynomial is arranged from lowest-degree term to highest-degree term with the rightmost bit assigned to the α^0 term. For $m = 4$, this approach yields:

$$1 + \alpha + \alpha^2 + \alpha^3 \leftrightarrow 1111$$
$$1 + \alpha \leftrightarrow 0011$$
$$1 \leftrightarrow 0001$$
$$\alpha + \alpha^3 \leftrightarrow 1010$$

This approach is used by Berlekamp.[6]

3. The polynomial is arranged from highest-degree term to lowest-degree term with the rightmost bit assigned to the α^0 term. For $m = 4$, this approach yields:

$$\alpha^3 + \alpha^2 + \alpha + 1 \leftrightarrow 1111$$
$$\alpha + 1 \leftrightarrow 0011$$
$$1 \leftrightarrow 0001$$
$$\alpha^3 + \alpha \leftrightarrow 1010$$

This approach is used by Peterson and Weldon.[7]

4. The polynomial is arranged from highest-degree term to lowest-degree term with the leftmost bit assigned to the α^0 term. For $m = 4$, this approach yields:

$$\alpha^3 + \alpha^2 + \alpha + 1 \leftrightarrow 1111$$
$$\alpha + 1 \leftrightarrow 1100$$
$$1 \leftrightarrow 1000$$
$$\alpha^3 + \alpha \leftrightarrow 0101$$

25

Although it is not as frequently used as approach 1, approach 3 is the most computationally convenient and is therefore used in the remainder of this book.

The Bottom Line

Recipe 2.1 can be used to construct the binary extension field $GF(2^m)$ provided that we have the following things:

☐ a primitive polynomial $p(x)$ of degree m with coefficients drawn from $GF(2)$.

☐ an ability to compute $xf(x)$ given $f(x)$ a polynomial in x.

☐ an ability to use the primitive polynomial to reduce the degree of $xf(x)$ whenever the result has a degree higher than $m - 1$.

6 E. R. Berlekamp: *Algebraic Coding Theory*, McGraw-Hill, New York, 1968.

7 W. W. Peterson and E. J. Weldon, Jr.: *Error-Correcting Codes*, 2nd ed., MIT Press, Cambridge, MA, 1972.

2.6 Computer Methods for Galois Fields

The Basic Idea

Our present goal is to implement Recipe 2.1 in software. This involves defining a format for storing coefficients, and designing software to generate the elements of $GF(2^m)$. The generation of these elements comprises two major sub-tasks:

1. Given a field element expressed as a polynomial $f(\alpha)$, compute $\alpha f(\alpha)$.

2. Use a primitive polynomial to reduce the degree of $\alpha f(\alpha)$ whenever the result of 1 has a degree higher than $m - 1$.

Programming Considerations Now let's examine the results of Example 2.4 with an eye towards computer realization of Recipe 2.1:

1. The first element is 0000.

2. The second element is 0001.

3. The third element can be obtained by shifting the second element one place to the left to obtain

$$\alpha^1 \leftrightarrow 0010$$

4. Likewise, the fourth and fifth element can be obtained by left-shifting the third and fourth elements to obtain:

$$\alpha^2 \leftrightarrow 0100$$
$$\alpha^3 \leftrightarrow 1000$$

5. If we try to obtain α^4 by a simple left shift of α^3, we will obtain the 5-bit value of 10000, which is incorrect. Instead, the sixth element α^4 is defined by the minimal polynomial selected in step 3 of Example 2.4:

$$\alpha^4 = \alpha + 1 \leftrightarrow 0011$$

6. The binary values for α^5 and α^6 can be obtained via left-shifting of α^4 and α^5:

$$\alpha^5 \leftrightarrow 0110$$
$$\alpha^6 \leftrightarrow 1100$$

7. If we shift the binary value for α^6 one bit to the left, we obtain 11000, which is a 5-bit value. The correct value for α^7 is 1011. The trick is to keep only the rightmost 4 bits of 11000 and add 0011. What is this really doing? It's simply making use of $\alpha^4 = \alpha + 1$. Removing the leftmost bit of 11000 corresponds to subtracting α^4 from the polynomial, and adding 0011 corresponds to adding $\alpha + 1$ to the polynomial.

These observations can be generalized into the following recipe which can be used directly to obtain the m-bit binary representations of the elements in GF(2^m).

Recipe 2.2 Constructing the binary extension field GF(2^m).

1. The first element is the all-zero value.

2. The second element has the least significant bit (LSB) set to 1, with all remaining bits set to 0.

3. Shift all bits one position to the left, shifting a zero into the LSB. If this shifting operation causes a 1 to be shifted out of the leftmost bit of the rightmost m bits, then mask away all but the rightmost m bits and exclusive-or (XOR) the result with the m-bit binary representation corresponding to α^m.

4. Repeat step 3 a total of $2^m - 2$ times.

2.6.1 Representing Field Elements

Elements of GF(2^m) can be represented in a computer in one of two different ways. Consider the extension field GF(2^4) shown in Table 2-13. The computer representation of each element can be either an integer $0, 1, 2,...,(2^m - 2)$ that equals the exponent of the element when expressed as a power of α, or the element can be expressed as an m-bit value directly corresponding to the m-tuple representation of the element. Each form has advantages and disadvantages. As shown previously, multiplication is very easy with the exponential form; the exponents are simply added:

$$\alpha^m \bullet \alpha^n = \alpha^{(m + n)}$$

However, addition is easier in the m-tuple form. In the approach used for manual computations, the multiplications are performed in exponential form, and additions are performed in m-tuple form. One form is easily converted to the other by referring to a table of the elements in the particular extension field of interest. This approach can also be implemented in a computer for relatively small values of m. Large values of m require large amounts of memory to hold the table of field elements. The table containing the elements for GF(2^m) will require 2^m locations with each location containing at least m bits. For $m = 16$, the table will need 131,072 bytes of memory. For some computer systems this might not be a problem; but for many small systems, an alternative approach might be needed. The ideal solution would be an algorithmic means for converting directly between the exponential form and m-tuple form without using a table lookup. Unfortunately, a simple direct conversion algorithm does not exist. However, there are several algorithmic conversion strategies that can be used in lieu of a memory-hogging lookup table.

Brute Force

A brute force approach would entail generating the desired field $GF(2^m)$ one element at a time without permanently saving the result. As each element is generated, it is examined to see if it is the element needed for the conversion being performed. Each element in succession is generated, examined, and discarded until the required element is obtained. Once the required element is obtained, the conversion is performed and the process is finished. This approach is very thrifty with memory, but **on average**, it will require 2^{m-1} iterations to reach the field element required for a conversion.

Modified Method

The brute force approach can be modified to reduce somewhat the number of iterations needed to find the required field element when converting from exponential form to m-tuple form. Simply generate the field elements using the methods of Section 2.5, but only store every k-th element. This will create an array of m-tuple values for elements α^j with $j = 1, 1 + k, 1 + 2k, 1 + 3k,...$ To convert an exponent, for example e, into m-tuple form we find the j-th element in the table such that $j \leq e < (j + k)$. Then, starting with the m-tuple representation for α^j, we use the method of Section 2.5 to generate the m-tuple representations for α^{j+1}, α^{j+2}, $\alpha^{j+3},...$ stopping when the element α^e is reached. For $m = 16$, the brute force method will take, on average, 32,768 iterations to perform the conversion. For $m = 16$ and $k = 1,024$, the modified method will take on average only 512 iterations to perform this conversion. For this particular case, the table needed for the brute force method will occupy only 128 bytes of memory. The speed-for-memory trade can be tailored to any particular computer system by an appropriate selection for k. Several different combinations of table size and iterations per conversion for the $m = 16$ case are listed in Table 2-14.

■ Table 2-14 Table sizes and iteration counts for converting forms of GF(2^k).

k	Table size (bytes)	Avg. iterations per conversion
1	131,072	0
2	65,536	1
4	32,768	2
8	16,384	4
16	8,192	8
32	4,096	16
64	2,048	32
128	1,024	64
256	512	128
512	256	256
1,024	128	512
2,048	64	1,024
4,096	32	2,048
8,192	16	4,096
16,384	8	8,192
32,768	4	16,384

Software Implementation

Listing 2-1 contains the BuildExtensionField() function that builds a table of field elements for small values of m. The required input alpha_m can be obtained by setting a primitive polynomial $p(\alpha)$ of degree m over GF(2) equal to zero and then solving for α^m. The variables Galois_Field[] and Num_Field_Elements are module variables that are shared with other functions in module gfield.c. This module contains a function GetAlphaM[] that accesses a data file primpoly.dat to obtain a value of α^m in the proper format to use as an input to BuildExtensionField(). As shown, BuildExtensionField() uses unsigned integers in order to represent field elements in m-tuple form for $m \leq 16$. Before BuildExtensionField() can be used to construct GF(2^m) for $m > 9$, the dimensions of Galois_Field[] must be increased to a value greater than or equal to 2^m. The Software Appendix discusses number of utility functions that are provided for performing arithmetic and format conversions on field elements.

Listing 2-1

```
static unsigned int Galois_Field[512];
static unsigned int Num_Field_Elements;
```

```
/*****************************/
/*                           */
/*   BuildExtensionField()   */
/*                           */
/*****************************/

void BuildExtensionField( int m, int alpha_m)
{
unsigned int shifted_out_mask, keeper_mask, i;

shifted_out_mask = IntPower(2, m);
keeper_mask = shifted_out_mask-1;
Galois_Field[0]=1;

for(i=1; i<keeper_mask; i++) {
  Galois_Field[i] = Galois_Field[i-1]<<1;
  if(Galois_Field[i] >= shifted_out_mask) {
    Galois_Field[i] &= keeper_mask;
    Galois_Field[i] ^= alpha_m;
    }
  }
Num_Field_Elements = IntPower(2,m);
return;
}
```

2.7 More About Polynomials

The Basic Idea

Polynomials with coefficients drawn from finite fields play an important role in the analysis and application of error correcting codes. We will frequently have need to represent various such polynomials in a "computer-friendly" form. Polynomials whose coefficients are drawn from a prime field GF(p) (see Section 2.2) are easy to deal with—use ordinary integer operations on the coefficients and reduce the result modulo p. Polynomials whose coefficients are drawn from an extension field GF(p^m) are a bit more difficult. This section explores several methods for representing such polynomials in software.

Assume that we have a polynomial in x with coefficients drawn from GF(2^m):

$$f(x) = \sum_{n=0}^{N} \beta_n x^n \qquad (2.4)$$

$$\beta_n \in \{0, 1, \alpha, \alpha^2, \alpha^3, ..., \alpha^{2^m-2}\}$$

How can we represent such a polynomial in software? We could store the coefficients of $f(x)$ in an array, for example beta[], with the coefficient for x^n stored in location n. However, we are faced with a choice between storing the coefficients in m-tuple form or as the exponents of their power form.

Before making a choice between the m-tuple or exponential representations, we should take a close look at the implications of option. The m-tuple representation has the following properties:

☐ Addition of two coefficients in m-tuple form is accomplished as the bit-by-bit exclusive-or (XOR) of the coefficients.

31

☐ Storage of the element 0 presents no special problems.

☐ Multiplication of two coefficients is not straightforward.

The exponential representation has the following properties:

☐ Multiplication of two coefficients is accomplished by adding the two exponents and reducing the sum modulo $(2^m - 1)$.

☐ The exponential form for the element 0 is $\alpha^{-\infty}$. This form might require special handling. (The value of 0 is reserved for representing α^0.) Two obvious candidates for representing $\alpha^{-\infty}$ are -1 and $2^m - 1$. Any positive value (including $2^m - 1$) is probably a bad idea because there might be times when we need these values to temporarily store the result of an operation before reducing the exponent modulo $(2^m - 1)$.

☐ Addition of two coefficients is not straightforward.

Back to our original problem; we wish to multiply a polynomial $p(x)$ by $(x + \alpha^k)$. Assume that m-bit binary representations of the coefficients for $p(x)$ are stored in an array beta[], with the coefficient for the x^n term stored in beta[n]. We further assume that the highest degree term in $p(x)$ is x^N (or, in other words, beta[i] = 0 for $i = N + 1, N + 2, N + 3,...$). We begin the design of our polynomial multiplication algorithm by noting that

$$(x + \alpha^k)\, p(x) = xp(x) + \alpha^k p(x)$$

Implementation of the term $xp(x)$ from this expansion is very easy—we simply shift the contents of location n into location $n + 1$ for $n = N, N - 1, N - 2,..., 0$. Implementation of the term $\alpha^k p(x)$ is straightforward; simply multiply each location of beta[] by α^k. Of course, this multiplication must be performed using the rules of multiplication defined for $GF(2^m)$, and the result must be added to the location-shifted version of beta[] using the rules of addition defined for $GF(2^m)$. The tricky part is to efficiently combine both operations so that the coefficient array can be updated "in place." The following fragment of C code makes use of function MultEx ponFieldElems( ) and AddExponFieldElems( ) to accomplish the desired arithmetic over $GF(2^m)$. Code to multiply a polynomial over $GF(2^m)$ by a binomial over $GF(2^m)$ is also incorporated in the Build OntoPolynomial( ) function presented in Section 2.9 for constructing minimal polynomials. In that code, the operations embedded within MultExponFieldElems( ) and AddExponFieldElems( ) are expanded inline to illustrate an alternative approach.

Program Fragment

```
/* The following fragment computes (x + alpha) * p(x)
where p(x)
is a polynomial in x of degree nn originally and of
degree nn+1 after the multiplication.  The coefficients
of p(x) are stored in beta[], and the values of these
coefficients are elements of the symbol field GF(2**m).
*/

for( n=nn; n>=0; n--){
    beta[n+1] = AddExponFieldElems( beta[n],
MultExponFieldElems(alpha, beta[n]));
    }
nn++;

/*  end of fragment  */
```

2.8 Cyclotomic Cosets

The Basic Idea

Cyclotomic cosets are an essential element in an algorithm for finding minimal polynomials which in turn play a crucial role in the construction of BCH codes and Reed-Solomon codes. This section defines cyclotomic cosets and develops a program for partitioning the elements of $GF(2^m)$ into cyclotomic cosets.

Assume that we are interested in the extension field $GF(2^m)$. As shown in Section 2.3, the nonzero elements of $GF(2^m)$ can be written in exponential form as $\alpha^0, \alpha^1, \alpha^2, ..., \alpha^{2m-2}$. This set of elements can be just as easily represented by the set of their exponents $\{0, 1, 2,..., 2^m - 2\}$. We could, if we wished, partition this set of exponents into subsets in any number of ways. For coding work, it turns out to be very useful to partition the set of exponents in one particular way. The partitioning technique is easier to understand if we begin with a specific example. Consider the set of exponents corresponding to the nonzero elements of $GF(2^4)$:

$$C = \{0, 1, 2,..., 14\}$$

The exponent 0 always goes into a subset by itself:

$$C_0 = \{0\}$$

Begin the next subset with the element 1:

$$C_1 = \{1,$$

To add another element to C_1, multiply the newest element by 2 and reduce the result modulo $2^4 - 1 = 15$:

$$C_1 = \{1, 2,$$

Continue adding elements in this fashion until the new element to be added is 1:

$$C_1 = \{ 1, 2$$
$$\downarrow$$
$$2 \times 2 = 4 \ (\mathrm{mod}15)$$
$$\downarrow$$
$$C_1 = \{ 1, 2, 4$$
$$\downarrow$$
$$2 \times 4 = 8 \ (\mathrm{mod}\ 15)$$
$$\downarrow$$
$$C_1 = \{ 1, 2, 4, 8$$
$$\downarrow$$
$$2 \times 8 = 1 \ (\mathrm{mod}\ 15)$$
$$C_1 = \{1, 2, 4, 8\}$$

The next subset is started with the smallest element in C that has not already been placed in a subset. In this particular example, we have

$$C_3 = \{3,$$

We continue adding elements as we did for C_1, but this time we stop when the new element to be added is 3:

$$C_3 = \{3$$
$$\downarrow$$
$$2 \times 3 = 6$$
$$\downarrow$$
$$C_3 = \{3, 6$$
$$\downarrow$$
$$2 \times 6 = 12 \ (\text{mod} 15)$$
$$\downarrow$$
$$C_3 = \{3, 6, 12$$
$$\downarrow$$
$$2 \times 12 = 9 \ (\text{mod } 15)$$
$$\downarrow$$
$$C_3 = \{3, 6, 12, 9$$
$$\downarrow$$
$$2 \times 9 = 3 \ (\text{mod } 15)$$
$$C_3 = \{3, 6, 12, 9\}$$

The next unused element is 5, so we have

$$C_5 = \{5$$
$$\downarrow$$
$$2 \times 5 = 10$$
$$\downarrow$$
$$C_5 = \{5, 10$$
$$\downarrow$$
$$2 \times 10 = 5 \ (\text{mod} 15)$$
$$C_5 = \{5, 10\}$$

Continuing in a similar fashion, we find

$$C_7 = \{7, 14, 13, 11\}$$

Now each element of C has been placed into one and only one subset:

$$C_0 = \{0\}$$
$$C_1 = \{1, 2, 4, 8\}$$
$$C_3 = \{3, 6, 12, 9\}$$
$$C_5 = \{5, 10\}$$
$$C_7 = \{7, 14, 13, 11\}$$

Subsets formed in this manner are called *cyclotomic cosets* mod $2^m - 1$. The concept of cyclotomic cosets can be generalized to any set of integers mod $p^m - 1$, with each coset consisting of

$$\{s, ps, p^2 s, p^3 s,..., p^{n-1} s\}$$

where each element is reduced modulo $p^m - 1$ and n is the smallest integer for which $p^n s \equiv s \pmod{p^n - 1}$.

Software Notes Listing 2-2 contains the BuildCyclotomicCosets() function that separates the integers $\{1, 2, ..., p^m - 2\}$ into cyclotomic cosets. The variables Coset_Array[][] and Num_Cosets are module variables that are shared with other functions in module gfield.c.

Listing 2-2

```
static exponent Coset_Array[32][MAX_ELEMENTS];
static int Num_Cosets;

/*******************************/
/*                             */
/*   BuildCyclotomicCosets     */
/*                             */
/*******************************/
void BuildCyclotomicCosets( int p,
                            int m)

/*--------------------------------------------------*/
/*  This routine partitions the integers            */
/*  {1,2,3,...((2**m)-2) } into cyclotomic           */
/*  cosets and places these cosets in the            */
/*  module variable Coset_Array[].  For cosets       */
/*  produced by this routine, the first element      */
/*  in each coset will be the smallest element       */
/*  in that coset.  The first elements will          */
/*  also be odd.  Therefore, without fear of         */
/*  conflict, we can place the coset with            */
/*  first element k into row (k+1)/2 of the          */
/*  module variable Coset_Array[].  The first        */
/*  location in each row (i.e. Coset_Array[k][0])    */
/*  contains the number of elements in that row's    */
/*  coset.                                           */

{
int modulus, i ,j, first_element,coset_index;
int working_element;
int element_count, total_count;
logical match;

modulus = IntPower(p,m)-1;
match = FALSE;
coset_index=0;
total_count = 0;
first_element=-1;
```

```
Coset_Array[1][0]=0;

do {
   /*  keep checking successive odd integers until
       one is found that is not already in one of
       the completed cosets. */
   do {
     first_element +=2;
     match = FALSE;
     for(i=1; i<=coset_index; i++) {
       for(j=1; j<=(signed)Coset_Array[i][0]; j++) {
         /*  if a candidate element is already
         in a coset, skip the row corresponding
         to this coset in Coset_Array[] and move
         on to the next odd candidate first element.*/

         if(first_element == (signed)Coset_Array[i][j]) {
           match = TRUE;
           Coset_Array[(first_element+1)/2][0]=0;
           i=coset_index+1;
           break;
           }
         }
       }
     }
   while(match);
   /* match remains FALSE if the element under
   consideration is suitable for use as the
   first element in a new coset. */
   coset_index++;

   /*  Generate all the elements of the next coset  */

   working_element=first_element;
   element_count=0;
   do {
     total_count++;
     element_count++;

Coset_Array[coset_index][element_count]=(exponent)working_element;
     working_element=(p*working_element) % modulus;
     }
   while (working_element != first_element);

   /*  Element zero holds the number of elements in this coset */

   Coset_Array[coset_index][0]=(exponent)element_count;
   }
while (total_count<(modulus-1));
Num_Cosets = coset_index;
return;
}
```

2.9 Finding Minimal Polynomials

For an element β of GF(2^m), the minimal polynomial $\phi(x)$ is the polynomial of smallest degree with coefficients drawn from GF(2) such that $\phi(\beta) = 0$. For example, the element α^5 from $GF(2^4)$ has as its minimal polynomial

$$\phi_5(x) = x^2 + x + 1$$

because

$$\begin{aligned}
\phi_5(\alpha^5) &= (\alpha^5)^2 + (\alpha^5) + 1 \\
&= \alpha^{10} + \alpha^5 + 1 \\
&= (\alpha^2 + \alpha + 1) + (\alpha^2 + \alpha) + 1 \\
&= 0
\end{aligned}$$

Defining and finding minimal polynomials is more than just an academic exercise—minimal polynomials are crucial to the construction of certain types of codes, such as BCH codes and Reed-Solomon codes.

Clearly, the minimal polynomial of the element 0 is always $\phi(x) = x$ for any extension field $GF(2^m)$. The minimal polynomial $\phi_k(x)$ for any nonzero element α^k can be obtained as

$$\phi^k(x) = \prod_{r=0}^{R-1} \left[x + (\alpha^k)^{2^r} \right]$$

where R is the smallest integer such that $(\alpha^k)^{2^R} = \alpha^k$.

Based upon the way the field $GF(2^m)$ is defined, it will always be the case that $(\alpha^k)^{2^m} = \alpha^k$. However, for some values of α^k, it might turn out that $(\alpha^k)^{2^R} = \alpha^k$ for some values of R less than m. Because $\alpha^{2^m-1} = 1$, we can conclude that

$$(\alpha^k)^{2^R} = \alpha^{k2^R} = \alpha^k$$

whenever

$$k2^R \equiv k \ (\text{modulo } 2^m - 1) \tag{2.5}$$

For any given k and m, it is a simple matter to find the smallest R for which Eq. 2.5 is satisfied by successively trying $R = 1$, $R = 2$, and so on up to $R = m$. (Remember that $R = m$ is guaranteed to work.) This leads us directly to the recipe for finding a minimal polynomial.

Recipe 2.3 Generating the minimal polynomial for element α^k of the field $GF(2^m)$.

1. Determine the smallest positive integer R for which

$$k2^R \equiv k \ (\text{modulo } 2^m - 1)$$

2. Using the value of R found in step 1, compute the minimal polynomial $\phi_k(x)$ of the element α^k as

$$\phi_k(x) = \prod_{r=0}^{R-1} \left(x + \alpha^{k2^r} \right) \tag{2.6}$$

Example 2.5

Find the minimal polynomial for each element in $GF(2^4)$.

1. $R = 1$ for α^0; $R = 2$ for α^5 and α^{10}; and $R = 4$ for all other elements.

2. Using Eq. 2.6, we obtain

$$\phi_0(x) = x + \alpha^0 = x + 1$$

$$\phi_1(x) = (x + \alpha)(x + \alpha^2)(x + \alpha^4)(x + \alpha^8)$$

$$= (x^2 + \alpha x + \alpha^2 x + \alpha^3)(x^2 + \alpha^8 x + \alpha^4 x + \alpha^{12})$$

$$= x^4 + \alpha^8 x^3 + \alpha^4 x^3 + \alpha^{12} x^2 + \alpha x^3 + \alpha^9 x^2$$

$$\quad + \alpha^5 x^2 + \alpha^{13} x + \alpha^2 x^3 + \alpha^{10} x^2 + \alpha^6 x^2$$

$$\quad + \alpha^{14} x + \alpha^3 x^2 + \alpha^{11} x + \alpha^7 x + \alpha^{15}$$

$$= x^4 + x^3(\alpha^8 + \alpha^4 + \alpha + \alpha^2)$$

$$\quad + x^2(\alpha^{12} + \alpha^9 + \alpha^5 + \alpha^{10} + \alpha^6 + \alpha^3)$$

$$\quad + x(\alpha^{13} + \alpha^{14} + \alpha^{11} + \alpha^7) + 1$$

$$= x^4 + x + 1$$

In a similar fashion, minimal polynomials can be obtained for α^2 through α^{14}:

$$\phi_2(x) = \phi_4(x) = \phi_8(x) = x^4 + x + 1$$

$$\phi_3(x) = \phi_6(x) = \phi_9(x) = \phi_{12}(x) = x^4 + x^3 + x^2 + x + 1$$

$$\phi_5(x) = \phi_{10}(x) = x^2 + x + 1$$

$$\phi_7(x) = \phi_{11}(x) = \phi_{13}(x) = \phi_{14}(x) = x^4 + x^3 + 1$$

Tables of minimal polynomials of elements in $GF(2^m)$ for $2 \le m \le 9$ are provided in Appendix.

Software Notes Listing 2-3 contains the BuildMinimalPolynomials() function from the module gfield.c. This function constructs the minimal polynomials for the extension field $GF(2^m)$ that resides in the module variable Galois_Field[]. The polynomial under construction is multiplied by successive binomial factors via calls to Build OntoPolynomial(). The variables Num_Field_Elements, Coset_Array[][], Minimal_Polynomial[], Degree_Of_Min_Poly[], and Num_Cosets are module variables that are shared with the other functions in module gfield.c.

Listing 2-3

```
static unsigned int Num_Field_Elements;
static exponent Coset_Array[32][MAX_ELEMENTS];
static word Minimal_Polynomial[MAX_ELEMENTS];
static int Degree_Of_Min_Poly[MAX_ELEMENTS];
static int Num_Cosets;
static int Degree_Of_Field;
```

```
/*******************************/
/*                             */
/*  BuildMinimalPolynomials()  */
/*                             */
/*******************************/

void BuildMinimalPolynomials(void)

/*-----------------------------------------------------*/
/*  all of the roots for any particular minimal        */
/*  polynomial will be contained within a single       */
/*  cyclotomic coset.  Therefore, this routine         */
/*  makes use of the module variable Coset_Array[][]   */
/*  and assumes that this array has been properly      */
/*  filled with each row containing the elements       */
/*  of one coset and the first location in each row    */
/*  containing the number of elements in the coset.    */
{
element poly_coeff[64];
element root;
int n, num_roots;
int max_degree,coset_index,first_index_in_poly_coeff;
logical high_degree_to_left;
word packed_min_poly;

/*-----------------------------------------------------*/
/* the coefficients of intermediate results will be    */
/* elements of GF(2**m), so the working polynomials    */
/*  are left in unpacked form.                         */

for(n=0;n<64;n++) poly_coeff[n]=0;
for( coset_index=1; coset_index<=Num_Cosets; coset_index++) {
  max_degree = 1;
  num_roots = Coset_Array[coset_index][0];
  poly_coeff[1]=1;
  poly_coeff[0] = ExponToTuple(Coset_Array[coset_index][1]);

  for( n=2; n<=num_roots; n++) {
    root = Coset_Array[coset_index][n];
    BuildOntoPolynomial(poly_coeff, &max_degree, root);
    printf("max_degree set to %d\n",max_degree);
    }

  /*-----------------------------------------------------*/
  /*  The coefficients of the final result will be       */
  /*  elements of GF(2), so we can pack the minimal      */
  /*  polynomial's coefficients into a single            */
  /*  binary word.                                       */
  high_degree_to_left = TRUE;
  first_index_in_poly_coeff = 0;
  PackBitsLong( poly_coeff,
          &packed_min_poly,
          max_degree + 1,
          first_index_in_poly_coeff,
          high_degree_to_left);
  /*----------------------------------------------------- */
  /* copy packed representation of minimal polynomial into */
  /* each array location corresponding to one of its roots. */
  /* The table so defined will be very useful for decoding. */
```

```
/* The table is held in the module variable minimalPolynomial[],*/
/* and routines external to this module can access the table */
/* using the function GetMinPoly().                           */
for (n=1; n<=num_roots; n++) {

Minimal_Polynomial[(Coset_Array[coset_index][n])]=packed_min_poly;
   Degree_Of_Min_Poly[(Coset_Array[coset_index][n])]=max_degree;
    printf("Degree_Of_Min_Poly[%d] set to %d\n",
          Coset_Array[coset_index][n], max_degree);
    }
  printf("\n %d) M%d = ",coset_index,
Coset_Array[coset_index][1]);
   for(n=max_degree; n>=0; n--) {
     printf("%d",poly_coeff[n]);
     }
   printf(" = %#lx",packed_min_poly);
   }
printf("\n");
return;
}
```

Block and Cyclic Codes

3.1 Introduction

A *block code* is a code that operates upon fixed-length blocks of information bits. (This is in contrast to convolutional codes that operate upon a continuous stream of information bits.) For ease of reference, the blocks of information bits are usually called *message blocks*—even though it might take many such blocks to transmit what we would normally think of as a "message" (a telegram, military orders, a computer-to-computer file transfer, etc.). As shown in Fig. 3-1, a block-code encoder takes a message block of k bits and produces an n-bit block, which is referred to as a *code word*, *code vector*, or *code block*. For any code, n will always be greater than k, so the encoder must supply $(n - k)$ additional bits. These additional bits are obtained by applying some particular set of rules to the bits in the k-bit message block. The exact nature of these rules is what defines a specific code: Hamming, Reed-Solomon, BCH, etc. The encoder might or might not produce a code word that explicitly contains the message block somewhere within the code word. Because each message block contains k bits, 2^k possible messages can be conveyed by each block. Each code word contains n bits, so there could be as many as 2^n distinct code words. However, block codes are designed to use only 2^k of the 2^n possible code words—one n-bit code word for each distinct k-bit message. Viewed another way, a block code uses n bits to transmit a message containing only k bits of useful information. Therefore, the code word contains some redundancy. It is this redundancy that is the source of the code's ability to detect and/or correct errors.

Assume that noise or other channel impairments corrupt the code word so that one of the unused n-bit patterns appears at the receiver. Because this is an unused or "illegal" pattern, the receiver "knows" that an error must have occurred. The trick is to devise a decoding scheme that will let the receiver decide which of the legal code words was actually transmitted. If too many bits within the block are corrupted, the receiver might make this decision in-

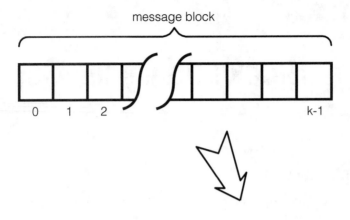

message block

0 1 2 k-1

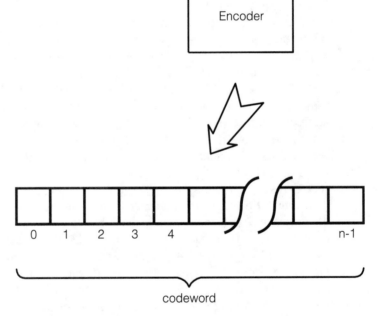

Encoder

0 1 2 3 4 n-1

codeword

■ **3-1** *Operation of a block encoder.*

correctly. In fact, if enough bits are corrupted, it is possible for a legal (but incorrect) code word to appear at the receiver. In such a situation, the receiver will not even be aware that an error has occurred. In general, the more redundancy there is in the code word [*i.e.*, the larger that $(n - k)$ is], the more bit errors the code will be able to detect and/or correct.

■ Table 3-1 An example of a simple block code.

Info block	Code word
0 0 0 0	0 0 0 0 0
0 0 0 1	0 0 0 1 1
0 0 1 0	0 0 1 0 1
0 0 1 1	0 0 1 1 0
0 1 0 0	0 1 0 0 1
0 1 0 1	0 1 0 1 0
0 1 1 0	0 1 1 0 0
0 1 1 1	0 1 1 1 1
1 0 0 0	1 0 0 0 1
1 0 0 1	1 0 0 1 0
1 0 1 0	1 0 1 0 0
1 0 1 1	1 0 1 1 1
1 1 0 0	1 1 0 0 0
1 1 0 1	1 1 0 1 1
1 1 1 0	1 1 1 0 1
1 1 1 1	1 1 1 1 0

Example 3.1 A parity scheme that appends one bit to each four-bit nibble in a computer memory is a simple example of a block code that is capable of detecting a single-bit error within each nibble. There are 4 bits in each message block and 5 bits in each code-word. Table 3-1 lists the message blocks and code words for such a scheme using an even-parity rule.

It is fairly common (but not universal) to use n to denote the total number of bits in a code word and to use k to denote the number of information bits. The difference $(n - k)$ is the number of parity bits in the code word. Clearly, there are 2^{n-k} different combinations of the $n - k$ parity bits. Each different k-bit message block will use only one of these 2^{n-k} possible combinations of message bits. [This is not to say that each different k-bit message block will use a different $(n-k)$-bit parity field. Many coding schemes assign the same parity field to two or more different k-bit message blocks. In fact, for codes in which $k > (n - k)$, individual parity fields must be assigned to more than one block because the number of different message blocks (2^k) will be greater than the number of different parity fields (2^{n-k}).]

A large part of coding theory is concerned with identifying "good" ways to assign particular parity fields to message blocks. In fact, the rules for assigning parity fields to message blocks are what really defines any particular code. Another large part of coding theory is concerned with rules for extracting the original message block from a (possibly corrupted) code block.

3.2 Linear Block Codes

Much of the work done in the area of block codes is confined to *linear* block codes, which are an important subset of all possible block codes. The usual definition for a linear block code is couched in terms of the mathematical structures defined in Chapter 2.

Definition 3.1 A binary block code having 2^k codewords (i.e., k information bits) and n-bits per codeword is a *linear* (n, k) code if and only if the 2^k codewords constitute a k-dimensional subspace of the vector space formed by all the n-tuples over the field GF(2).

This definition can be recast in a more practical, user-oriented form as follows:

Definition 3.2 A binary block code is *linear* if and only if the modulo-2 sum of any two codewords is also a codeword.

3.2.1 Encoding for Linear Block Codes

By Definition 3.1, the code $\mathbf{C}$ is a k-dimensional vector space; therefore, $\mathbf{C}$ can be completely defined by k linearly independent codewords. This can be exploited to form the *generator matrix* $\mathbf{G}$ such that the encoding operation is represented mathematically as

$$\mathbf{v} = \mathbf{u} \bullet \mathbf{G}$$

where $\mathbf{v}$ is a vector of the encoded bits, $\mathbf{u}$ is a vector of k information bits, and $\mathbf{G}$ is the *generator matrix*. Take a more detailed look at the specific case of a block code having a total of n bits per code word, with k of these being information bits and $(n - k)$ being parity bits. For such a code, $\mathbf{v}$ will have n bits, $\mathbf{u}$ will have k bits, and $\mathbf{G}$ will have k rows and n columns:

$$\mathbf{u} = [u_0 \ u_1 \ u_2 \ \cdots \ u_{k-1}],$$
$$\mathbf{v} = [v_0 \ v_1 \ v_2 \ \cdots \ v_{n-1}],$$
$$\mathbf{G} = \begin{bmatrix} g_{00} & g_{01} & g_{02} & \cdots & g_{0,n-1} \\ g_{10} & g_{11} & g_{12} & \cdots & g_{1,n-1} \\ \vdots & & \ddots & & \vdots \\ g_{k-1,0} & g_{k-1,1} & g_{k-1,2} & \cdots & g_{k-1,n-1} \end{bmatrix}$$

By the standard "row-by-column" definition of matrix multiplication, each component of **v** is given by

$$v_i = \sum_{j=0}^{k-1} u_j g_{ji} \tag{3.1}$$

where the summation is performed using binary addition.

Programming Considerations The following fragment of C code implements Eq. 3.1 for each of the n components in **v**:

```
/****************************************/

for( i=0; i<n; i++){
    v[i] =0;
    for( j=0; j<k; j++ ) {
        v[i] = v[i] ^ (u[j] & g[j][i]);
        }
    }
/*   end of fragment   */
/****************************************/
```

If the codewords within a code can be partitioned into two parts, with one part consisting of the k-bit information word and the other part consisting of the $n - k$ parity digits, the code is said to be a *systematic* code or a code in *systematic form*. The alternative, in which the message digits do not appear directly in the codeword, is described as a *nonsystematic* code. If a code is in systematic form, its generator matrix will be separable into two submatrices such that one of the submatrices is the $k \times k$ identity matrix. This is the part of the generator matrix that "copies" the k-digit message word into the code word. The remaining submatrix calculates the parity digits as linear functions of the message digits. Notationally:

$$\mathbf{G} = [\mathbf{I}_k \; \mathbf{P}]$$

where

$\mathbf{I}_k = k \times k$

$\mathbf{P} = k \times (n - k)$ matrix of parity-check coefficients

Any linear code that is not in systematic form can be converted into systematic form by performing elementary row operations and column transpositions on its generator matrix. Any changes to **G** via elementary row operations leaves the total set of codewords unchanged—even though it changes the particular one-to-one correspondence between k-bit information vectors and n-bit codewords.

Example 3.2 Consider the (7, 4, $d = 3$) code shown in Table 3-2 and having a generator matrix given by

$$\mathbf{G} = \begin{bmatrix} 1 & 1 & 0 & 0 & 0 & 1 & 0 \\ 0 & 1 & 1 & 0 & 0 & 0 & 1 \\ 0 & 0 & 1 & 1 & 1 & 0 & 1 \\ 0 & 0 & 0 & 1 & 0 & 1 & 1 \end{bmatrix} \qquad (3.2)$$

Perform row operations as necessary to obtain an equivalent (7, 4, $d = 3$) code in systematic form.

■ Table 3-2 A block code in nonsystematic form for Example 3.2.

Info block	Code word
0 0 0 0	0 0 0 0 0 0 0
0 0 0 1	0 0 0 1 0 1 1
0 0 1 0	0 0 1 1 1 0 1
0 0 1 1	0 0 1 0 1 1 0
0 1 0 0	0 1 1 0 0 0 1
0 1 0 1	0 1 1 1 0 1 0
0 1 1 0	0 1 0 1 1 0 0
0 1 1 1	0 1 0 0 1 1 1
1 0 0 0	1 1 0 0 0 1 0
1 0 0 1	1 1 0 1 0 0 1
1 0 1 0	1 1 1 1 1 1 1
1 0 1 1	1 1 1 0 1 0 0
1 1 0 0	1 0 1 0 0 1 1
1 1 0 1	1 0 1 1 0 0 0
1 1 1 0	1 0 0 1 1 1 0
1 1 1 1	1 0 0 0 1 0 1

Solution Starting with the matrix given in Table 3-2, perform the following row operations:

1. Add row 4 to row 3.
2. Add the new row 3 to row 2.
3. Add the new row 2 to row 1.

The resulting matrix is

$$\mathbf{G} = \begin{bmatrix} 1 & 0 & 0 & 0 & 1 & 0 & 1 \\ 0 & 1 & 0 & 0 & 1 & 1 & 1 \\ 0 & 0 & 1 & 0 & 1 & 1 & 0 \\ 0 & 0 & 0 & 1 & 0 & 1 & 1 \end{bmatrix} \qquad (3.3)$$

The corresponding code is listed in Table 3-3. Notice that the first four columns of the new matrix form a submatrix that is an identity matrix. Comparison of the two tables reveals that **G** from Eq. 3.2 and its systematic form from Eq. 3.3 generate the same sixteen 7-tuples.

■ **Table 3-3 Block code in systematic form for Example 3.2.**

Info block	Check block
0 0 0 0	0 0 0
0 0 0 1	0 1 1
0 0 1 0	1 1 0
0 0 1 1	1 0 1
0 1 0 0	1 1 1
0 1 0 1	1 0 0
0 1 1 0	0 0 1
0 1 1 1	0 1 0
1 0 0 0	1 0 1
1 0 0 1	1 1 0
1 0 1 0	0 1 1
1 0 1 1	0 0 0
1 1 0 0	0 1 0
1 1 0 1	0 0 1
1 1 1 0	1 0 0
1 1 1 1	1 1 1

3.2.2 Constructing the Generator Matrix

How do we go about obtaining a generator matrix? As we will see in later chapters, some codes (*e.g.,* Hamming codes) are defined directly in terms of their generator matrices, while most cyclic codes are defined in terms of their generator *polynomial.*

3.2.3 Parity-Check Matrix

For any linear code **C**, it is possible to construct a matrix **H** such that

$$\mathbf{v} \bullet \mathbf{H}^{\perp} = 0 \qquad (3.4)$$

if and only if the n-tuple **v** is a codeword in **C**. This matrix **H** is usually called the *parity-check* matrix for code **C**. If a particular n-tuple **r** is received, Eq. 3.4 can be used to determine whether or not $\mathbf{r}$ is a legal codeword in $\mathbf{C}$. The decoder computes the *syndrome* $\mathbf{s}$ as

$$\mathbf{s} = \mathbf{r} \cdot \mathbf{H}^{\perp} \tag{3.5}$$

The matrix H has $n - k$ rows and n columns, and the vector $\mathbf{r}$ has n components. Therefore, based on the definition of matrix multiplication ("row by column"), we know that the syndrome $\mathbf{s}$ will be a vector having $n - k$ components

$$\mathbf{s} = [s_0, s_1, ..., s_{n-k-1}]$$

If $\mathbf{s} = [0, 0,..., 0]$, then $\mathbf{r}$ is a codeword in $\mathbf{C}$. If $\mathbf{s} \neq [0, 0,..., 0]$, then $\mathbf{r}$ is not a codeword in $\mathbf{C}$. The study of decoding is primarily concerned with ways that nonzero values of the syndrome $\mathbf{s}$ can be exploited to actually locate and correct errors within the received word $\mathbf{r}$.

The parity-check matrix $\mathbf{H}$ can be obtained from the generator matrix $\mathbf{G}$ via

$$\mathbf{G} = [\mathbf{I}_k \ \mathbf{P}]$$
$$\mathbf{H} = [-\mathbf{P}^{\perp} \ \mathbf{I}_{n-k}] \tag{3.6}$$

where $\mathbf{P}^{\perp}$ is the transpose of $\mathbf{P}$. The minus sign on $\mathbf{P}^{\perp}$ is moot in the binary case because $-1 \equiv 1$, but it is included here for consistency in nonbinary applications.

Example 3.3 Using the generator matrix for the (7, 4) systematic code found in Example 3.2, find the parity-check matrix $\mathbf{H}$.

Solution We first partition $\mathbf{G}$ into the form $[\mathbf{I}_k \mathbf{P}]$ as

$$\mathbf{G} = \begin{bmatrix} 1 & 0 & 0 & 0 & 1 & 0 & 1 \\ 0 & 1 & 0 & 0 & 1 & 1 & 1 \\ 0 & 0 & 1 & 0 & 1 & 1 & 0 \\ 0 & 0 & 0 & 1 & 0 & 1 & 1 \end{bmatrix}$$

Making use of Eq. 3.6, we obtain $\mathbf{H}$ as

$$\mathbf{H} = \begin{bmatrix} 1 & 1 & 1 & 0 & 1 & 0 & 0 \\ 0 & 1 & 1 & 1 & 0 & 1 & 0 \\ 1 & 1 & 0 & 1 & 0 & 0 & 1 \end{bmatrix}$$

Let's take a closer look at what's really happening in Eq. 3.5. Based on the partitioning of $\mathbf{H}$ given in Eq. 3.6, we can write $\mathbf{H}^{\perp}$ and $\mathbf{s}$ as

$$\mathbf{H}^{\perp} = \begin{bmatrix} \mathbf{P} \\ \mathbf{I}_{n-k} \end{bmatrix}$$

48

$$\mathbf{s} = \mathbf{r} \bullet \mathbf{H}^{\perp} = [r_0, r_1, ..., r_{n-1}] \begin{bmatrix} \mathbf{P} \\ \mathbf{I}_{n-k} \end{bmatrix}$$

$$= [r_0, r_1, ..., r_{n-1}] \begin{bmatrix} p_{00} & p_{01} & \cdots & p_{0,n-k-1} \\ p_{10} & p_{11} & & p_{1,n-k-1} \\ \vdots & & \ddots & \vdots \\ p_{k-1,0} & p_{k-1,1} & \cdots & p_{k-1,n-k-1} \\ 1 & 0 & 0 \cdots & 0 \\ 0 & 1 & & 0 \\ \vdots & & \ddots & \vdots \\ 0 & 0 & 0 \cdots & 1 \end{bmatrix}$$

The ith component of $\mathbf{s}$ is obtained as the vector multiplication of $\mathbf{r}$ with column i of $\mathbf{H}^{\perp}$

$$s_i = r_0 p_{0i} + r_1 p_{1i} + ... + r_{k-1} p_{k-1,i} + r_{k+i}$$

$$= r_{k+i} + \sum_{j=0}^{k-1} r_j p_{ji}$$

The first term, r_{k+i}, is simply the ith parity bit as received at the decoder. The summation is a recomputation of this same parity bit using the parity-check coefficients p and the k received message bits r_0 through r_{k-1}. If the received parity and recomputed parity are equal, the sum of these two will be zero.

Definition 3.3 Let $\mathbf{H}$ be the parity-check matrix of code $\mathbf{C}$. If we use $\mathbf{H}$ as a generator matrix instead of a parity-check matrix, the resulting code $\mathbf{C}_d$ is called the *dual code* of $\mathbf{C}$.

3.3 Cyclic Codes

A special class of linear block codes, called *cyclic* codes, possess mathematical properties that make them easier to encode and decode than linear codes in general. The two most widely used types of linear block codes—BCH codes and Reed-Solomon codes—are cyclic codes.

Given any n-tuple $\mathbf{v} = [v_0, v_1, ..., v_{n-1}]$, it is possible to cyclically shift the n-tuple so that each component moves one place to the right and the rightmost component becomes the leftmost component.

$$\mathbf{v}_{\text{shifted}} = [v_{n-1}, v_0, v_1, ..., v_{n-2}]$$

This concept can be extended to include shifts of multiple places and shifts in either direction.

Definition 3.4 An (n, k) linear block code **C** is a *cyclic code* if every cyclic shift of a codeword in **C** is also a codeword in **C**.

Example 3.4 Consider the code from Example 3.2. We can sort the codewords into groups as shown in Table 3.4 such that the codewords within each group are cyclic shifts of each other. Each group contains all possible shifts of the basic pattern corresponding to that group.

■ Table 3-4
Codewords for linear (7,4) code sorted into groups such that all codewords within a group are cyclic shifts of each other.

Info block	Check block
0 0 0 0	0 0 0
0 0 0 1	0 1 1
0 0 1 0	1 1 0
0 1 0 1	1 0 0
1 0 1 1	0 0 0
0 1 1 0	0 0 1
1 1 0 0	0 1 0
1 0 0 0	1 0 1
0 1 0 0	1 1 1
1 0 0 1	1 1 0
0 0 1 1	1 0 1
0 1 1 1	0 1 0
1 1 1 0	1 0 0
1 1 0 1	0 0 1
1 0 1 0	0 1 1
1 1 1 1	1 1 1

By definition, if a particular n-tuple is a legal codeword for a cyclic code **C**, then all cyclic shifts of this n-tuple are also legal codewords. Conversely, if a particular n-tuple is not a legal codeword for a cyclic code **C**, then none of the cyclic shifts of the n-tuple can be a legal codeword.

Result 3.1 In a cyclic code **C**, the nonzero code polynomial of minimum degree is unique. (Recall from Section 2.5 that each element of an extension field can be represented in polynomial form. Because each codeword in a block code can be represented as an element of an extension field, it follows that each codeword can be represented by a polynomial.)

Definition 3.5 For a cyclic code **C**, the nonzero code polynomial of minimum degree is called the *generator polynomial* of the code. It is customary to denote the generator polynomial by $\mathbf{g}(x)$.

Result 3.2 If the generator polynomial of a cyclic code is given by

$$\mathbf{g}(x) = x^r + g_{r-1}x^{r-1} + g_{r-2}x^{r-2} + \dots + g_1 x + g_0$$

the constant term g_0 will always equal 1.

Result 3.3 In an (n, k) cyclic code **C**, the degree of the generator polynomial is $n - k$, *i.e.*,

$$\mathbf{g}(x) = x^{n-k} + g_{n-k-1}x^{n-k-1} + g_{n-k-2}x^{n-k-2} + \dots + g_2 x^2 + g_1 x + 1$$

Result 3.4 Let $\mathbf{g}(x)$ be the generator polynomial of a cyclic code **C**. A polynomial of degree $n - 1$ or less with binary coefficients is a code polynomial of **C** if and only if the polynomial is divisible by $\mathbf{g}(x)$.

Result 3.5 If $\mathbf{g}(x)$ is the generator polynomial of an (n, k) cyclic code, $\mathbf{g}(x)$ is a factor of $x^n + 1$.

Result 3.6 If $\mathbf{g}(x)$ is a polynomial of degree $n - k$ and is a factor of $x^n + 1$, then $\mathbf{g}(x)$ is the generator polynomial for some (n, k) cyclic code.

This last result says that $\mathbf{g}(x)$ will generate some (n, k) cyclic code—it does not say that every $\mathbf{g}(x)$ that is a factor of $x^n + 1$ produces a good code.

3.4 Manual Encoding Methods for Cyclic Codes

In our explorations of specific codes, it will be useful to have a manual pencil-and-paper method for generating codewords corresponding to specific information fields. Recipe 3.1 provides such a method.

Recipe 3.1 Manual encoding for cyclic codes

1. Obtain the generator polynomial $g(x)$. For a code with k information bits and n total bits per code word, the generator polynomial will be of degree $n - k$.

2. Form the polynomial representation of the k-bit information sequence.

3. Multiply this polynomial by x^{n-k}.

4. Divide this product by $g(x)$. This division yields a quotient $q(x)$ and a remainder $r(x)$. This remainder is the polynomial representation of the check bits.

5. Form the code word by appending the check bits to the information bits.

Example 3.5 Given the generator polynomial $\mathbf{g}(x) = x^3 + x^2 + 1$, use Recipe 3.1 to generate the 7-bit codeword for the 4-bit information sequence 0100.

1. We have $\mathbf{g}(x) = x^3 + x^2 + 1$, with $n = 7$ and $k = 4$.

2. The information sequence 0100 corresponds to the polynomial $\mathbf{u}(x) = x^2$.

3. Multiply $\mathbf{u}(x)$ by x^{n-k},
$$x^{n-k}\mathbf{u}(x) = x^3 x^2 = x^5$$

4. Divide x^5 by $x^3 + x^2 + 1$

$$
\begin{array}{r}
x^2 + x + 1 \\
\hline
x^3 + x^2 + 1 \overline{)\, x^5 } \\
x^5 + x^4 + x^2 \\
\hline
x^4 + x^2 \\
x^4 + x^3 + x \\
\hline
x^3 + x^2 + x \\
x^3 + x^2 + 1 \\
\hline
x + 1
\end{array}
$$

The remainder shown corresponds to the 3-bit parity sequence 011.

5. The code word is formed by appending the parity bits to the information bits:

$$\underbrace{0100}_{\text{info}} \quad \overbrace{011}^{\text{parity}}$$

3.5 Modifications to Cyclic Codes

The various codes produced in the quest for good error-correction properties might not always be the right size to fit conveniently into real-world applications. This section will present several techniques that are available for adjusting the sizes of n, k, and d for existing cyclic codes.

3.5.1 Extended Codes

One simple way to modify a code is to *extend* the code by including additional check digits. In the binary case, this is frequently accomplished by appending an overall parity bit to an (n, k, d) code to obtain an $(n + 1, k, d')$ code. If the original code's minimum distance d is odd, and the parity rule for the extra bit results in an overall even parity, the extended code's minimum distance will be $d' = d + 1$.

Example 3.6 Consider the $(7, 4, d = 3)$ code listed in Table 3-5. An overall even parity bit can be appended to yield the $(8, 4, d = 4)$ extended code listed in Table 3-6.

■ **Table 3-5**
A (7, 4, d = 3)
block code for
modification examples.

Info block	Check block	Weight
0 0 0 0	0 0 0	0
0 0 0 1	1 0 1	3
0 0 1 0	1 1 1	4
0 0 1 1	0 1 0	3
0 1 0 0	0 1 1	3
0 1 0 1	1 1 0	4
0 1 1 0	1 0 0	3
0 1 1 1	0 0 1	4
1 0 0 0	1 1 0	3
1 0 0 1	0 1 1	4
1 0 1 0	0 0 1	3
1 0 1 1	1 0 0	4
1 1 0 0	1 0 1	4
1 1 0 1	0 0 0	3
1 1 1 0	0 1 0	4
1 1 1 1	1 1 1	7

53

■ Table 3-6 The (8, 4, d = 4)
extended code resulting
from adding an overall even
parity bit to the code of Table 3-5.

Info block	Check block	Weight
0 0 0 0	0 0 0 0	0
0 0 0 1	1 0 1 1	4
0 0 1 0	1 1 1 0	4
0 0 1 1	0 1 0 1	4
0 1 0 0	0 1 1 1	4
0 1 0 1	1 1 0 0	4
0 1 1 0	1 0 0 1	4
0 1 1 1	0 0 1 0	4
1 0 0 0	1 1 0 1	4
1 0 0 1	0 1 1 0	4
1 0 1 0	0 0 1 1	4
1 0 1 1	1 0 0 0	4
1 1 0 0	1 0 1 0	4
1 1 0 1	0 0 0 1	4
1 1 1 0	0 1 0 0	4
1 1 1 1	1 1 1 1	8

3.5.2 Punctured Codes

Puncturing a code decreases the number of check bits and the total number of bits, while leaving the number of information bits
and the number of codewords unchanged. This is accomplished by
simply deleting bit positions from the check portion of the codeword structure. In most cases, puncturing will also decrease the
minimum distance of a code. Puncturing can be viewed as the inverse of extending a code. Suppose that we are starting with a
code that has an even minimum distance, such as the one in Table
3-6. Such a code can detect up to $d/2$ errors per block, but its error-correction capability is no better than a code that has an odd
minimum distance $d' = d - 1$. Therefore, in situations where we are
only concerned with error correction, it might be desirable to decrease the length of a code by eliminating one of the check bits,
provided that this can be done without decreasing the number of
errors that can be corrected.

3.5.3 Expurgated Codes

Expurgation of a code decreases the number of information bits k
without changing the total number of bits n. The number of code-

words is always 2^k, so each time k is decreased by 1, the number of codewords is halved. The most common approach to expurgation deletes all the odd-weight codewords. Deletion of odd-weight codewords is equivalent to replacing the original code's generator polynomial $\mathbf{g}(x)$ with $\mathbf{g}'(x) = (x + 1)\,\mathbf{g}(x)$. When applied to the (7, 4, d = 3) code of Table 3-5, this approach yields the (6, 2, d = 4) code of Table 3-7. In the new code, the boundary between information bits and check bits has been shifted to take one bit away from the information portion and prepend it to the check portion. The remaining information bits cover all 2^3 possible 3-bit messages. Will it always work out this way, or for some expurgated codes will some possible messages be covered twice while others are not covered? In Table 3-5, the codewords can be paired up so that within each pair the first 3 of the 4 information bits are identical. In the general case, it will always be possible to pair up the codewords so that within each pair, the first $k - 1$ of the k information bits are identical. For any code not already having $(x + 1)$ as a factor of its generator polynomial, one codeword of each pair will have odd weight and one codeword will have even weight. Therefore, when we eliminate all the odd-weight codewords, the even-weight members of each pair will survive and thereby "cover" all 2^{k-1} possible combinations of $k - 1$ information bits.

■ Table 3-7
The (7, 3, d = 4) block code
resulting from expurgation.

Info block	Check block	Weight
0 0 0	0 0 0 0	0
0 0 1	0 1 1 1	4
0 1 0	1 1 1 0	4
0 1 1	1 0 0 1	4
1 0 0	1 0 1 1	4
1 0 1	1 1 0 0	4
1 1 0	0 1 0 1	4
1 1 1	0 0 1 0	4

3.5.4 Augmented Codes

Augmentation can be viewed as the inverse of expurgation. Any code having $(x + 1)$ as a factor of its generator polynomial $\mathbf{g}(x)$ can be augmented by replacing the original generator polynomial with $\mathbf{g}'(x) = \mathbf{g}(x)/(x + 1)$. In most practical applications, augmentation has little to offer.

3.5.5 Shortened Codes

In any linear block code, half of the codewords will have 0 in the most significant bit (MSB) of the information segment and half will have 1. We can shorten an (n, k, d) code by discarding all codewords with 1 in the MSB of the information portion and then eliminating this bit from the codeword structure to obtain an $(n - 1, k - 1, d')$ code. The minimum distance of the shortened code will be at least as large as the minimum distance of the original code (*i.e.*, $d' \geq d$). This shortening process can be extended to multiple bits by retaining only those codewords having all zeros in the z most significant bit positions and then eliminating these z positions from the codeword structure to obtain an $(n - z, k - z, d_z)$ code.

Example 3.7 Consider the $(7, 3, d = 4)$ code shown in Table 3-7. If we retain only those codewords with 0 in the MSB of the information portion, and then eliminate this bit position, we obtain the $(6, 2, d = 4)$ code shown in Table 3-8.

■ **Table 3-8 The (6, 2, d = 4) block code resulting from shortening the (7, 4, d = 4) code of Table 3-7.**

Info block	Check block
0 0	0 0 0 0
0 1	0 1 1 1
1 0	1 1 1 0
1 1	1 0 0 1

3.5.6 Lengthened Codes

Lengthening can be viewed as the inverse of shortening, but it is rarely used in practical applications.

3.6 Hamming Codes

Hamming codes were one of the first codes developed specifically for error correction. Like many binary codes, the lengths of Hamming codes are constrained to be of the form

$$n = 2^m - 1$$

However, unlike other codes of similar length, there is a much tighter coupling between the number of check bits and the length of the code—for $n = 2^m - 1$, the number of check bits is m and the number of information bits is $k = 2^m - m - 1$. The minimum distance is fixed at 3; consequently, the number of correctable errors per block is $t = 1$. The systematic form of the generator matrix is given by

$$\mathbf{G} = [\mathbf{I}_k \quad \mathbf{Q}^\perp]$$

where

$\mathbf{I}_k = k \times k$ identity matrix

$\mathbf{Q}^\perp = k \times m$ matrix whose rows are the k m-tuples of weight 2 or larger.

The corresponding parity-check matrix is given by

$$\mathbf{H} = [\mathbf{Q} \quad \mathbf{I}_m]$$

Example 3.8 Construct the generator and parity-check matrices for a Hamming code with $m = 4$.

Solution For $m = 4$, we have $n = 2^4 - 1 = 15$ and $k = n - m = 11$. One possible realization of the generator matrix is

$$\mathbf{G} = \begin{bmatrix}
1 & 0 & 0 & 0 & 0 & 0 & 0 & 0 & 0 & 0 & & 0 & 0 & 1 & 1 \\
0 & 1 & 0 & \ldots & & & & & & & 0 & 0 & 1 & 0 & 1 \\
0 & 0 & 1 & 0 & \ldots & & & & & & 0 & 0 & 1 & 1 & 0 \\
0 & \ldots & 0 & 1 & 0 & \ldots & & & & & 0 & 0 & 1 & 1 & 1 \\
0 & \ldots & & 0 & 1 & 0 & \ldots & & & & 0 & 1 & 0 & 0 & 1 \\
0 & \ldots & & & 0 & 1 & 0 & \ldots & & & 0 & 1 & 0 & 1 & 0 \\
0 & \ldots & & & & 0 & 1 & 0 & \ldots & & 0 & 1 & 0 & 1 & 1 \\
0 & \ldots & & & & & 0 & 1 & 0 & \ldots & 0 & 1 & 1 & 0 & 0 \\
0 & \ldots & & & & & & 0 & 1 & 0 & 0 & 1 & 1 & 0 & 1 \\
0 & \ldots & & & & & & & 0 & 1 & 0 & 1 & 1 & 1 & 0 \\
0 & 0 & 0 & 0 & 0 & 0 & 0 & 0 & 0 & 0 & 1 & 1 & 1 & 1 & 1
\end{bmatrix} \quad (3.7)$$

The corresponding parity-check matrix is given by

$$\mathbf{H} = \begin{bmatrix}
0 & 0 & 0 & 0 & 1 & 1 & 1 & 1 & 1 & 1 & 1 & 1 & 0 & 0 & 0 \\
0 & 1 & 1 & 1 & 0 & 0 & 0 & 1 & 1 & 1 & 1 & 0 & 1 & 0 & 0 \\
1 & 0 & 1 & 1 & 0 & 1 & 1 & 0 & 0 & 1 & 1 & 0 & 0 & 1 & 0 \\
1 & 1 & 0 & 1 & 1 & 0 & 1 & 0 & 1 & 0 & 1 & 0 & 0 & 0 & 1
\end{bmatrix} \quad (3.8)$$

It would be impractical to list the $2^{11} = 2{,}048$ possible codewords for this code.

3.6.1 Shortened Hamming Codes

It is possible to shorten an $(n = 2^m - 1, k = 2^m - m - 1, d = 3)$ Hamming code to produce an $(n = 2^{m-1}, k = 2^{m-1} - m, d = 4)$ code. Simply delete all columns of even weight from the submatrix $\mathbf{Q}$ and all rows of even weight from the submatrix $\mathbf{Q}^\perp$. A minimum distance of 4 is not large enough to correct more than one error, but it will allow simultaneous correction of one error and detection of two errors if the following decoding strategy is followed:

1. Compute the syndrome from the received word $\mathbf{r}$.

2. If the syndrome is zero, assume that no errors have occurred.

3. If the syndrome is nonzero and has odd weight, assume that a single error has occurred. Perform correction based on the single-error pattern corresponding to the syndrome observed.

4. If the syndrome is nonzero and has even weight, assume that an uncorrectable error pattern has been detected.

Example 3.9 Shorten the $(15, 11, d = 3)$ Hamming code of Example 3.8 to create an $(8, 4, d = 4)$ code.

Solution Delete all rows of even weight from the $\mathbf{Q}^\perp$ submatrix shown in Eq. 3.7 and combine with a 4-×-4 identity matrix to obtain

$$\mathbf{G} = \begin{bmatrix} 1\ 0\ 0\ 0 & 0\ 1\ 1\ 1 \\ 0\ 1\ 0\ 0 & 1\ 0\ 1\ 1 \\ 0\ 0\ 1\ 0 & 1\ 1\ 0\ 1 \\ 0\ 0\ 0\ 1 & 1\ 1\ 1\ 0 \end{bmatrix}$$

Delete all columns of even weight from the $\mathbf{Q}$ submatrix shown in Eq. 3.8 and combine with a 4-×-4 identity matrix to obtain

$$\mathbf{H} = \begin{bmatrix} 0\ 1\ 1\ 1 & 1\ 0\ 0\ 0 \\ 1\ 0\ 1\ 1 & 0\ 1\ 0\ 0 \\ 1\ 1\ 0\ 1 & 0\ 0\ 1\ 0 \\ 1\ 1\ 1\ 0 & 0\ 0\ 0\ 1 \end{bmatrix}$$

The resulting code is listed in Table 3-9.

■ Table 3-9 The (8, 4, d = 4) code obtained by shortening the (15, 11, d = 3) Hamming code of Example 3.8.

Info bits	Check bits	Weight
0 0 0 0	0 0 0 0	0
0 0 0 1	1 1 1 0	4
0 0 1 0	1 1 0 1	4
0 0 1 1	0 0 1 1	4
0 1 0 0	1 0 1 1	4
0 1 0 1	0 1 0 1	4
0 1 1 0	0 1 1 0	4
0 1 1 1	1 0 0 0	4
1 0 0 0	0 1 1 1	4
1 0 0 1	1 0 0 1	4
1 0 1 0	1 0 1 0	4
1 0 1 1	0 1 0 0	4
1 1 0 0	1 1 0 0	4
1 1 0 1	0 0 1 0	4
1 1 1 0	0 0 0 1	4
1 1 1 1	1 1 1 1	8

BCH and
Reed-Solomon Codes

4

One very often-used type of linear cyclic block codes are the Bose-Chaudhuri-Hocquenghem or BCH codes discovered by R. C. Bose and D. K. Ray-Chaudhuri[1] and independently by A. Hocquenghem.[2] BCH codes are used extensively in a variety of practical applications; their popularity is caused in large part by the existence of computationally efficient and easily implemented decoding procedures, which are presented in Chapter 5. BCH codes can be binary with symbols 0 and 1 drawn from GF(2), or they can be nonbinary with symbols 0, 1, 2, ..., $p^m - 1$ drawn from GF(p^m). Unless explicitly identified as nonbinary, the term *BCH code* used by itself almost always refers to binary BCH codes. One particular type of nonbinary BCH code, called the Reed-Solomon code, is examined at length in Section 4.3.

4.1 BCH Codes

Coding theory provides a result, sometimes called the *BCH bound*, which forms the basis for construction of BCH codes.

Result 4.1 *BCH bound.* If a linear cyclic code is constructed such that

☐ each codeword contains n bits

☐ β is an element of order n from the extension field GF(2^m)

☐ the code's generator polynomial $\mathbf{g}(x)$ has included among its roots $(\delta - 1)$ consecutive powers of β

then

☐ the code is guaranteed to have a minimum distance of δ or greater.

---

1 R. C. Bose and D. K. Ray-Chaudhuri: "On a Class of Error Correcting Binary Group Codes," *Inf. Control*, 3, pp. 68-79, March 1960.

2 A. Hocquenghem: "Codes corecteurs d'erreurs," *Chiffres*, 2, pp. 147-156, 1959.

Notes

1. A proof of the BCH bound can be found in MacWilliams and Sloane,[3] Lin and Costello,[4] or Berlekamp.[5]

2. The result as stated above requires that the generator polynomial have $(\delta - 1)$ consecutive powers of β **included among its roots**. The generator polynomial can (and usually will) have other roots in addition to the $(\delta - 1)$ consecutive powers of β.

3. BCH codes are constructed by selecting a desired minimum distance δ and creating a generator polynomial that has $(\delta - 1)$ consecutive powers of a field element β as roots. The actual minimum distance of the code might turn out to be greater than δ, and therefore δ is usually called the *design distance* to distinguish it from the actual distance, which is usually denoted by d.

4.1.1 Genesis of the Codes

BCH codes are the result of a straightforward approach for creating a generator polynomial having the sequence of roots that the BCH bound says are needed to guarantee some desired minimum distance δ:

1. Assume that we wish to create a code capable of correcting up to τ errors per block. In order to correct τ errors, the code must have a minimum distance of at least $\delta = 2\tau + 1$.

2. According to the BCH bound, one way to guarantee that the code has a minimum distance of at least δ is to construct the generator polynomial $\mathbf{g}(x)$ in such a way that included among the roots of $\mathbf{g}(x)$ are $\delta - 1$ or 2τ consecutive powers of β. We can identify these "required" roots as:

$$\beta^{b+1}, \beta^{b+2}, ..., \beta^{b+2\tau} \qquad (4.1)$$

3. A generator polynomial having the roots listed in (4.1) can be obtained as

$$\mathbf{g}(x) = (x + \beta^{b+1})(x + \beta^{b+2})(x + \beta^{b+3})......(x + \beta^{b+\delta-1}), p(x) \qquad (4.2)$$

In addition to one factor for each of the "required" roots, the righthand side of Equation 4.2 includes a polynomial, $p(x)$. This polynomial includes additional roots that are needed to ensure that each coefficient of $\mathbf{g}(x)$ is either 0 or 1. When all of

3 F. J. MacWilliams and J. J. A. Sloane: *The Theory of Error-Correcting Codes,* North Holland, Amsterdam, 1977.

4 S. Lin and D. J. Costello: *Error Control Coding: Fundamentals and Applications,* Prentice-Hall, Englewood Cliffs, NJ, 1983.

5 E. R. Berlekamp: *Algebraic Coding Theory,* McGraw-Hill, New York, 1968.

62

the terms of the form $(x + \beta^r)$ are multiplied together, the result will have coefficients that are, in general, elements of the **extension** field GF(2^m). The polynomial $p(x)$ in Equation 4.1 includes factors corresponding to additional roots that are needed to force the coefficients of $\mathbf{g}(x)$ to be elements of the **prime** field GF(2). The additional factors needed to make up $p(x)$ can be determined using minimal polynomials.

4. For each of the "required" roots β^r, there is a minimal polynomial $M^{(r)}(x)$ that has β^r as a root [i.e., $M^{(r)}(\beta^r) = 0$], and has **binary** coefficients drawn from GF(2).

5. A generator polynomial that includes all the required roots and has binary coefficients can be obtained as the least common multiple (LCM) of all the minimal polynomials of the "required" roots:

$$\mathbf{g}(x) = \text{LCM}\{M^{(b+1)}(x), M^{(b+2)}(x), ..., M^{(b+\delta-1)}(x)\}$$

Often, several different roots will have the same minimal polynomial, which needs to be included in $\mathbf{g}(x)$ once and only once. Obtaining $\mathbf{g}(x)$ as the least common multiple of all the "required" minimal polynomials ensures that each distinct polynomial is included in $\mathbf{g}(x)$ once and only once. Two unequal minimal polynomials will never share a common root, so the LCM of two unequal minimal polynomials is the product of the two.

4.1.2 Types of BCH Codes

There are several varieties of BCH codes based upon different ways in which the $(\delta - 1)$ consecutive powers of β are selected for constructing the code's generator polynomial.

☐ If β is a primitive element of GF(2^m), the resulting BCH code is called a *primitive* BCH code and each codeword will contain $2^m - 1$ bits.

☐ If β is a nonprimitive element of GF(2^m), the resulting BCH code is called a *nonprimitive* BCH code, and the length of each codeword is equal to the order of the element β.

☐ If in Eq. 4.2, $b = 0$ the first of the $(\delta - 1)$ powers of β is $\beta^1 = \beta$, and the resulting BCH code is called a *narrow-sense* BCH code.

☐ If $b \neq 0$, the resulting BCH code is called a *wide-sense* BCH code.

4.1.3 Critical Features

☐ BCH codes are linear cyclic codes; therefore, they exhibit all of the properties of linear cyclic codes presented in Chapter 4.

Most notably:

~If r is a code word, then all cyclic permutations of r are code
words.

~Any specific BCH code is completely defined by its generator
polynomial, $\mathbf{g}(x)$.

☐ Specific binary BCH codes are denoted by ordered pairs
(n, k) or sometimes by ordered triples (n, k, t) or (n, k, d),
where

n = total number of bits in each codeword

k = number of information bits in each codeword

t = maximum number of errors that can be corrected in any

d = distance of the code

☐ The total number of words in the code is 2^k.

☐ For **primitive** BCH codes, the number of bits n is restricted
to be $2^m - 1$, where m is a positive integer.
~Legal code lengths are 3, 7, 15, 31, 63, 127, 255, etc.

☐ For **nonprimitive** BCH codes, the number of bits n is
restricted to be a value that divides $2^m - 1$ for some integer m.
Because $2^m - 1$ will always be odd, n must always be odd.
~Example: For $m = 6$, we find $(2^m - 1) = 63 = 7 \bullet 3 \bullet 3$;
therefore, the only values of n that divide 63 are 3, 7, 9,
and 21.

☐ For some values of t, it is not possible to design a code that
corrects exactly t errors per block.

☐ Example: Given a length $n = 63$, it is possible to design
primitive narrow-sense BCH codes for $t = 7$ and $t = 10$; but it
is not possible to design such a code for $t = 8$ or $t = 9$. If τ is
specified as 8 or 9 in the following design recipes, the recipes
will automatically produce the code for $t = 10$. In general, if τ
is specified as the desired value for t in either of the recipes in
this section, the recipe will automatically yield the code
having the smallest t that equals or exceeds τ. This will satisfy
the specified error-correction requirements, but having to
"trade-up" to the higher value of t causes more of the bits
within a code block to be "wasted" for check bits that provide
additional correction capability that we really didn't ask for.
Sections 5.4 and 5.5 present several ways to work around this
limitation. Values of n, k, and t for primitive narrow-sense
BCH codes are listed in Table 4-1.

■ Table 4-1 Combinations of *n*, *k*, and *t* for primitive BCH codes.

n	k	t	n	k	t	n	k	t
7	4	1	127	36	15	255	45	43
15	11	1		29	21		37	45
	7	2		22	23		29	47
	5	3		15	27		21	55
31	26	1		8	31		13	59
	21	2	255	247	1		9	63
	16	3		239	2	511	502	1
	11	5		231	3		493	2
	6	7		223	4		484	3
63	57	1		215	5		475	4
	51	2		207	6		466	5
	45	3		199	7		457	6
	39	4		191	8		448	7
	36	5		187	9		439	8
	30	6		179	10		430	9
	24	7		171	11		421	10
	18	10		163	12		412	11
	16	11		155	13		403	12
	10	13		147	14		394	13
	7	15		139	15		385	14
127	120	1		131	18		376	15
	113	2		123	19		367	16
	106	3		115	21		358	18
	99	4		107	22		349	19
	92	5		99	23		340	20
	85	6		91	25		331	21
	78	7		87	26		322	22
	71	9		79	27		313	23
	64	10		71	29		304	25
	57	11		63	30		295	26
	50	13		55	31		286	27
	43	14		47	42		277	28
511	268	29	1,023	1,003	2	1,023	688	36
	259	30		993	3		678	37
	250	31		983	4		668	38
	241	36		973	5		658	39
	238	37		963	6		648	41
	229	38		953	7		638	42
	220	39		943	8		628	43
	211	41		933	9		618	44
	202	42		923	10		608	45
	193	43		913	11		598	46
	184	45		903	12		588	47
	175	46		893	13		578	49

n	k	t	n	k	t	n	k	t
	166	47		883	14		573	50
	157	51		873	15		563	51
	148	53		863	16		553	52
	139	54		858	17		543	53
	130	55		848	18		533	54
	121	58		838	19		523	55
	112	59		828	20		513	57
	103	61		818	21		503	58
	94	62		808	22		493	59
	85	63		798	23		483	60
	76	85		788	24		473	61
	67	87		778	25		463	62
	58	91		768	26		453	63
	49	93		758	27		443	73
	40	95		748	28		433	74
	31	109		738	29		423	75
	28	111		728	30		413	77
	19	119		718	31		403	78
	10	121		708	34		393	79
1,023	1,013	1		698	35		383	82
1,023	378	83	1,023	238	109	1,023	121	171
	368	85		228	110		111	173
	358	86		218	111		101	175
	348	87		208	115		91	181
	338	89		203	117		86	183
	328	90		193	118		76	187
	318	91		183	119		66	189
	308	93		173	122		56	191
	298	94		163	123		46	219
	288	95		153	125		36	223
	278	102		143	126		26	239
	268	103		133	127		16	147
	258	106		123	170		11	255
	248	107						

☐ Examination of the way that cyclotomic cosets are generated for fields GF(2^m) (See Section 2.9) reveals that the minimal polynomials for the first τ odd powers of β will include among their roots the first τ nonzero even powers of β as well. Therefore, the set of minimal polynomials for the first τ odd powers of β will be identical to the set of minimal polynomials for the first 2τ (nonzero) integer powers of β. This means that for binary primitive narrow-sense BCH codes, only the first τ odd powers of β have to be explicitly considered in the construction of $\mathbf{g}(x)$—the first τ even powers will be automatically included.

☐ The relationships between n, k, $\mathbf{g}(x)$, and GF(2^m) are depicted in Fig. 4-1.

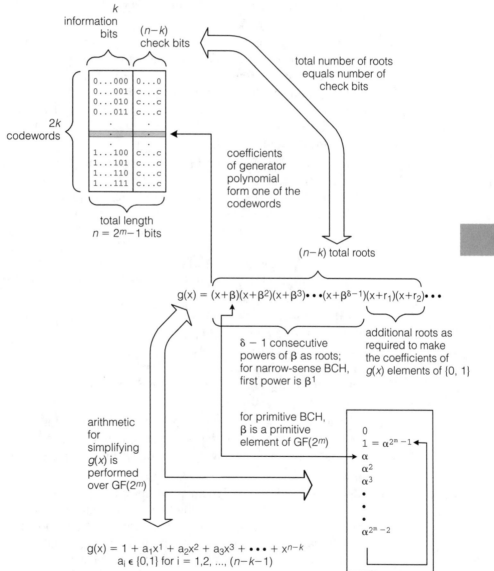

■ **4-1** *Relationships between n, k, **g**(x), and GF(2^m) for primitive narrow-sense BCH codes.*

4.1.4 Constructing the Generator Polynomial

Recipe 4.1 can be used to manually construct the generator polynomials for binary primitive narrow-sense BCH codes.

Recipe 4.1 *Constructing the generator polynomial for a primitive narrow-sense BCH code. (Table-driven approach for small n.)*

1. Select the desired values for m and n where the code length is $n = 2^m - 1$.

2. Select the value for τ which is the desired maximum number of errors to be corrected in each block of n bits.
   - ☐ The code will actually be capable of correcting t errors per block, and the design procedure guarantees that $t \geq \tau$.
   - ☐ Related to τ and t are the quantities $\delta = 2\tau + 1$ and $d = 2t + 1$, where δ is the *design distance* of the code and d is the *actual* minimum distance of the code.

3. Select an integer j between 1 and $2^m - 2$ such that j and $2^m - 1$ have no factors in common.
   - ☐ Example: For $m = 5$, $2^m - 1 = 31$, which is a prime. Therefore, j can be any value between 1 and 30 and not have any factors in common with 31.
   - ☐ Example: For $m = 4$, we find $2^m - 1 = 15$, which has factors 3 and 5. Therefore, j must be selected from {1, 2, 4, 7, 8, 11, 13, 14} because the remaining integers (3, 5, 6, 9, 10, and 12) contain factors of 3 or 5.

4. For the value of τ selected in step 2 and the value of j selected in step 3, find the first τ odd multiples of j: Whenever one of the calculated values exceeds $2^m - 2$, reduce the value modulo $(2^m - 1)$ to obtain a value between 1 and $2^m - 2$. The generator polynomial being constructed will have a root β^r for each value of $r = j, 3j, 5j, ..., (2\tau - 1)j$.

5. For each root β^r identified by the multiples of j found in step 4, select from the appropriate table in Appendix A, the minimal polynomial having β^r as a root. Often, a single minimal polynomial will have several of the β^r as roots.

6. Obtain the generator polynomial by multiplying together all the minimal polynomials selected in step 5. Each selected minimal polynomial is to be included in the multiplication only once, regardless of how many different β^r are roots of the polynomial. The multiplication of the polynomials is conducted using the methods of Section 2.5.

Example 4.1 Use Recipe 4.1 to construct the generator polynomial for a BCH code of length $n = 15$ with $\tau = 3$. Then apply the methods of Section 3.4 (**Manual Encoding for Cyclic Codes**) to obtain all of the codewords in this code.

1. For $n = 2^m - 1 = 15$, we find $m = 4$.

2. As given, $\tau = 3$.

3. Because $2^m - 1 = 15 = 3 \bullet 5$, and {3, 5, 6, 9, 10, 12} are multiplies of 3 and/or 5, the value for j must be selected from {1, 2, 4, 7, 8, 11, 13, 14}. Let's arbitrarily select $j = 1$.

4. The first 3 odd multiples of j are $j = 1$, $3j = 3$, and $5j = 5$.

5. From Table 4-3, the minimal polynomials for α^1, α^3, α^5 are

$$M^{(1)}(x) = x^4 + x + 1$$
$$M^{(3)}(x) = x^4 + x^3 + x^2 + x + 1$$
$$M^{(5)}(x) = x^2 + x + 1$$

6. We find the generator polynomial by multiplying $M^{(1)}(x)$, $M^{(3)}(X)$, and $M^{(5)}(x)$.

$$\mathbf{g}(x) = M^{(1)}(x); M^{(3)}(x); M^{(5)}(x)$$
$$= x^{10} + x^8 + x^5 + x^4 + x^2 + x + 1$$

7. Because the degree of $\mathbf{g}(x)$ is 10, the number of information bits in each codeword will be $k = n - 10 = 5$. Using $\mathbf{g}(x)$ from step 6 in conjunction with the CyclicEncoder() function provided in Listing 5-1, we find the 32-word BCH (15, 5, $t = 3$) code listed in Table 4-2. Examination of this table reveals that the actual minimum distance is 7, so the code is indeed a t = 3 code. (Remember, $d = 2t + 1$).

The BCH codes listed in most coding books are the codes generated when $j = 1$ is selected in step 3 of Recipe 4.1. However, according to the recipe, it is possible to generate codes by selecting $j \neq 1$. Consider the following example.

Example 4.2 Repeat Example 4.1, but select $j = 7$ instead of $j = 1$ in step 3.

1. $m = 4$.

2. $\tau = 3$.

3. $j = 7$.

4. The first 3 odd multiples of j are

$$j = 7$$
$$3j = 21, 21 \bmod 15 = 6$$
$$5j = 35, 35 \bmod 15 = 5$$

Info bits	Check bits
0 0 0 0 0	0 0 0 0 0 0 0 0 0 0
0 0 0 0 1	0 1 0 0 1 1 0 1 1 1
0 0 0 1 0	1 0 0 1 1 0 1 1 1 0
0 0 0 1 1	1 1 0 1 0 1 1 0 0 1
0 0 1 0 0	0 1 1 1 1 0 1 0 1 1
0 0 1 0 1	0 0 1 1 0 1 1 1 0 0
0 0 1 1 0	1 1 1 0 0 0 0 1 0 1
0 0 1 1 1	1 0 1 0 1 1 0 0 1 0
0 1 0 0 0	1 1 1 1 0 1 0 1 1 0
0 1 0 0 1	1 0 1 1 1 0 0 0 0 1
0 1 0 1 0	0 1 1 0 1 1 1 0 0 0
0 1 0 1 1	0 0 1 0 0 0 1 1 1 1
0 1 1 0 0	1 0 0 0 1 1 1 1 0 1
0 1 1 0 1	1 1 0 0 0 0 1 0 1 0
0 1 1 1 0	0 0 0 1 0 1 0 0 1 1
0 1 1 1 1	0 1 0 1 1 0 0 1 0 0
1 0 0 0 0	1 0 1 0 0 1 1 0 1 1
1 0 0 0 1	1 1 1 0 1 0 1 1 0 0
1 0 0 1 0	0 0 1 1 1 1 0 1 0 1
1 0 0 1 1	0 1 1 1 0 0 0 0 1 0
1 0 1 0 0	1 1 0 1 1 1 0 0 0 0
1 0 1 0 1	1 0 0 1 0 0 0 1 1 1
1 0 1 1 0	0 1 0 0 0 1 1 1 1 0
1 0 1 1 1	0 0 0 0 1 0 1 0 0 1
1 1 0 0 0	0 1 0 1 0 0 1 1 0 1
1 1 0 0 1	0 0 0 1 1 1 1 0 1 0
1 1 0 1 0	1 1 0 0 1 0 0 0 1 1
1 1 0 1 1	1 0 0 0 0 1 0 1 0 0
1 1 1 0 0	0 0 1 0 1 0 0 1 1 0
1 1 1 0 1	0 1 1 0 0 1 0 0 0 1
1 1 1 1 0	1 0 1 1 0 0 1 0 0 0
1 1 1 1 1	1 1 1 1 1 1 1 1 1 1

5. From Table A-3, the minimal polynomials for $i = 5, 6, 7$ are
$$M^{(5)}(x) = x^2 + x + 1$$
$$M^{(3)}(x) = x^4 + x^3 + x^2 + x + 1$$
$$M^{(7)}(x) = x^4 + x^3 + 1$$

6. We find the generator polynomial by multiplying $M^{(3)}(x)$, $M^{(5)}(x)$, and $M^{(7)}(x)$
$$\mathbf{g}(x) = M^{(3)}(x)M^{(5)}(x)M^{(7)}(x)$$
$$= x^{10} + x^9 + x^8 + x^6 + x^5 + x^2 + 1$$

It turns out that Examples 4.1 and 4.2 do not really generate different codes even though it appears as though they do. The generator polynomials can be written in vector form as

$$x^{10} + x^8 + x^5 + x^4 + x^2 + x + 1 \rightarrow (10100110111) \qquad (4.3)$$

$$x^{10} + x^9 + x^8 + x^6 + x^5 + x^2 + 1 \rightarrow (11101100101) \qquad (4.4)$$

The representation in Eq. 4.4 is simply the representation of Eq. 4.3 in bit-reversed order. It turns out that the code constructed using the generator polynomial of Example 4.2 is just the code constructed using the generator polynomial of Example 4.1 in bit-reversed order. The two codes are completely equivalent in terms of their distances, correction capabilities, and weight distributions. (When two different codes are equivalent in this way, the codes are referred to as *isomorphs* of each other.) If we were to repeat the examples for j = 2, 4, 8, 11, 13, 14, we would find that the result of Example 4.1 is obtained for j = 1, 2, 4, 8 and the result of Example 4.2 is obtained for j = 7, 11, 13, and 14.

4.1.5 Algorithmic Approach

As demonstrated by Examples 4.1 and 4.2, Recipe 4.1 is straightforward in its application, but the computations can become tedious for large values of n. Furthermore, Recipe 4.1 requires tables of minimal polynomials. In the remainder of this section, we will develop computer software for finding the generator polynomial of a binary primitive narrow-sense BCH code. This software will generate its own minimal polynomials as they are needed. As a first step, let's revise Recipe 4.1 to eliminate the use of tables for finding minimal polynomials.

Recipe 4.2 *Constructing the generator polynomial for a primitive narrow-sense BCH code. (Algorithmic approach.)*

1. Select the desired values for m and n where the code length is $n = 2^m - 1$.
2. Select the value for τ, which is the desired maximum number of errors to be corrected in each block of n bits.
3. Select an integer j between 1 and $2^m - 2$ such that j and $2^m - 1$ have no factors in common. [This step essentially selects from (2^m) a primitive element β where $\beta = \alpha^j$.]
4. For the value of τ selected in step 2 and the value of j selected in step 3, find the first τ odd multiples of j:

$$j, 3j, 5j, ..., (2\tau - 1)j$$

Whenever one of the calculated values exceeds $2^m - 2$, the value must be reduced modulo $(2^m - 1)$. This step finds the first τ odd powers of β, but these powers are expressed in terms of exponents of α where the roots are $\beta = \alpha^j$, $\beta^3 = \alpha^{3j}, \ldots, \beta^{(2\tau - 1)} = \alpha^{(2\tau - 1)j}$. Within the field $GF(2^m)$, we know that $\alpha^{(2^m - 1)} = 1 = \alpha^0$, so when a multiple of j (i.e., an exponent of α) exceeds $2^m - 2$, it can (and should) be reduced modulo $(2^m - 1)$ to obtain a value between 0 and $2^m - 2$.

5. Decompose $2^m - 1$ into cyclotomic cosets using the methods of Section 3.15.

6. Determine which of the cosets found in step 5 is "required." If a desired root of $\mathbf{g}(x)$ is $\beta^r = \alpha^{rj}$, then the cyclotomic coset containing the value rj is a "required" coset.

7. From each "required" coset, generate the corresponding "required" minimal polynomial. For a coset with elements $\{r_1, r_2, \ldots, r_m\}$, the corresponding minimal polynomial $M(x)$ can be obtained as

$$M(x) = (x + r_1)(x + r_2)\ldots(x + r_m)$$

8. Obtain the desired generator polynomial as the product of all "required" minimal polynomials.

Example 4.3 Use Recipe 4.2 to construct the generator polynomial for a BCH code of length $n = 15$ with $\tau = 3$.

Solution

1. For $n = 2^m - 1 = 15$, we find $m = 4$.

2. As given, $\tau = 3$.

3. Because $2^m - 1 = 15 = 3 \cdot 5$, and $\{3, 5, 6, 9, 10, 12\}$ are multiples of 3 and/or 5, the value for j must be selected from $(1, 2, 4, 7, 8, 11, 13,$ and 14. Let's arbitrarily select $j = 1$.

4. The first 3 odd multiples of j are $j = 1$, $3j = 3$, and $5j = 5$.

5. The cyclotomic cosets for $n = 15$ are

$$C_1 = \{1, 2, 4, 8\} \quad C_3 = \{3, 6, 12, 9\}$$
$$C_5 = \{5, 10\} \quad C_7 = \{7, 14, 13, 11\}$$

6. The "required" cosets are C_1, C_3, and C_5; because $1 \in C_1$, $3 \in C_3$, and $5 \in C_5$. Coset C_7 is not required because none of the multiples of j from step 4 are contained in C_7.

7. The corresponding "required" minimal polynomials are

$$M^{(1)}(x) = (x + \alpha)(x + \alpha^2)(x + \alpha^4)(x + \alpha^8)$$
$$= x^4 + x + 1$$

$$M^{(3)}(x) = (x + \alpha^3)(x + \alpha^6)(x + \alpha^{12})(x + \alpha^9)$$
$$= x^4 + x^3 + x^2 + x + 1$$
$$M^{(5)}(x) = (x + \alpha^5)(x + \alpha^{10})$$
$$= x^2 + x + 1$$

8. We find the generator polynomial by multiplying $M^{(1)}(x)$, $M^{(3)}(x)$, and $M^{(5)}(x)$.
$$\mathbf{g}(x) = M^{(1)}(x)M^{(3)}(x)M^{(5)}(x)$$
$$= x^{10} + x^8 + x^5 + x^4 + x^2 + x + 1$$

This result agrees with the result of Example 4.1.

4.1.6 Computer Approach

We now present Recipe 4.3, which is used as the basis for designing a computer program, BuildBchCode(), that will be able to construct the generator polynomial for any primitive narrow-sense BCH code. This recipe embodies the basic approach of Recipe 4.2, but some of the operations have been shuffled around a bit to make things more computer-friendly. Some of the steps have been annotated with ideas regarding ways to actually implement the steps in **C** code.

Recipe 4.3 *Constructing the generator polynomial for a primitive narrow-sense BCH code. (Computer-friendly approach.)*

1. Select the desired extension field $GF(2^m)$. This choice can be specified as an integer value of m plus the specific primitive polynomial used to generate the extension field. As discussed in Section 2.4, the coefficients of this polynomial can be represented as the vector form of field element α^m.

2. Specify the desired number of correctable errors per block as τ.

3. Select a primitive element β from $GF(2^m)$. This choice can be specified as an integer value for j where $\beta = \alpha^j$.

4. Set $\mathbf{g}(x) = 1$.

5. For odd values of i from 1 through $2\tau - 1$, perform the following: [Note: Each value of i corresponds to one of the odd powers of β to be included among the roots of $\mathbf{g}(x)$.]

   A. Compute $k = ij \bmod n$. Subsequent steps will cause the field element, α^k, to be incorporated as a root of $\mathbf{g}(x)$.

   B. If α^k has already been "used," set i to the next odd value, compute a new k, and check α^k again. This step is repeated until an "unused" α^k is found. [By the time this step is ready to include α^k as a root in $\mathbf{g}(x)$, step 5C might have already "used" α^k by causing it to be included in $\mathbf{g}(x)$ as part of the minimal polynomial for some other root. The check performed in this step ensures that no root appears more than once $\mathbf{g}(x)$. For how α^k gets marked as "used," see step 5Civ.]

C. For values of r from 0 through $m - 1$, perform the following: [This loop finds all the elements of the cyclotomic coset that contains k, and then multiplies $\mathbf{g}(x)$ by the minimal polynomial that corresponds to this coset. This multiplication is performed one root at a time in step 5Cv. For $n = 2^m - 1$, there can be at most m elements in any coset, hence the indexing of r from 0 to $m - 1$.]

   i. Compute $\lambda = k2^r$.

   ii. Reduce λ modulo $(2^m - 1)$.

   iii. If $\lambda = k$ and $r \neq 0$, all the elements for the current coset have been found. If this is the case, set i to the next odd value and go to step 5A. Otherwise, continue with step 5Civ.

   iv. Mark α^k as "used." This marking could be accomplished in several different ways. In one approach, an array (for example, alpha_used[]) is created with one location for each possible value of α^k. The contents of location k is set TRUE to mark α^k as "used." This approach requires alpha_used[] to have 2^m locations, and for large m, the memory requirements can become prohibitive. A second approach is to maintain a list of "used" α^k values. This approach requires more time to check each new value of α^k against the "used" list, but the memory requirements are quite reasonable. For a BCH code designed to correct τ errors, the generator polynomial will have at most $m\tau$ roots. For $m = 20$ and $\tau = 10$, the second approach requires 200 locations instead of the 1,048,576 locations required by the first approach.

   v. Multiply $\mathbf{g}(x)$ by $(x + \alpha^\lambda)$. The degree of $\mathbf{g}(x)$ will increase by one each time this step is executed.

   vi. Once step 5 has been completed for all values of i from 1 through $2\tau - 1$, the resulting $\mathbf{g}(x)$ will be equal to the desired BCH generator polynomial. As $\mathbf{g}(x)$ is accumulated in step 5Cv, the coefficients will be elements of either GF(2) or GF(2^m) represented in vector form. Whenever only a portion of a minimal polynomial's factors have been multiplied into $\mathbf{g}(x)$, the coefficients will be elements of GF(2^m). Whenever $\mathbf{g}(x)$ contains only factors of complete minimal polynomials, the coefficients will be elements of GF(2) [which, by the way, are also elements of GF(2^m)]. The conversion from GF(2^m) to GF(2) happens automatically because the computer's vector representations for 0 and 1 are the same in both GF(2) and GF(2^m).

Recipe 4.3 maps directly into the function BuildBchCode(), which is provided in Listing 4-1. A flowchart of this recipe is provided in Fig. 4-2.

Example 4.4 Use Recipe 4.3 to construct the generator polynomial for a primitive narrow-sense BCH code of length $n = 15$ with $\tau = 3$. Use $\beta = \alpha^7$ as the primitive element for step 2.

Solution

1. For $n = 15 = 2^4 - 1$, we must use $m = 4$.
2. As given, $\tau = 3$.
3. As given, $\beta = \alpha^7$, so $j = 7$.
4. $\mathbf{g}(x) = 1$
5. $i = 1$
   A. $k = ij \bmod 15 = 7$
   B. α^7 has not been used
   C. $r = 0$
      i. $\lambda = (7)(2^0) = 7$
      ii. $\lambda = 7$
      iii. continue
      iv. mark α^7 as used
      v. $\mathbf{g}(x) = x + \alpha^7$
   C. $r = 1$
      i. $\lambda = 14$
      ii. $\lambda = 14$
      iii. continue
      iv. mark α^{14} as used
      v. $\mathbf{g}(x) = (x + \alpha^7)(x + \alpha^{14}) = x^2 + \alpha x + \alpha^6$
   C. $r = 2$
      i. $\lambda = 28$
      ii. $\lambda = 13$
      iii. continue
      iv. mark α^{13} as used
      v. $\mathbf{g}(x) = (x^2 + \alpha x + \alpha^6)(x + \alpha^{13})$
      $\qquad = x^3 + \alpha^{12}x^2 + \alpha^8 x + \alpha^4$
   C. $r = 3$
      i. $\lambda = 56$
      ii. $\lambda = 11$

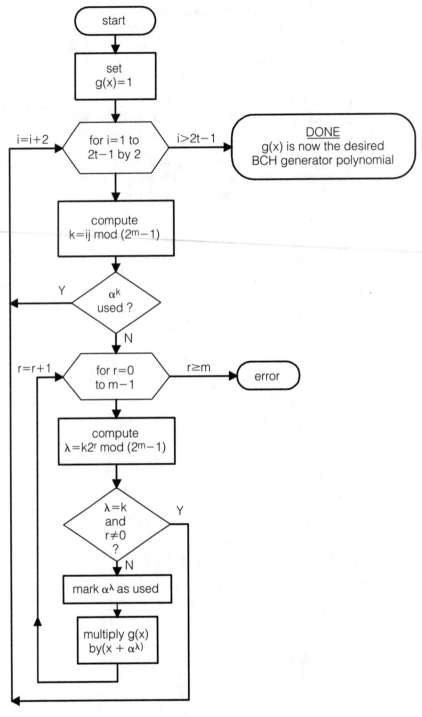

■ **4-2** *Flowchart for Recipe 4.3.*

iii. continue

iv. mark α^{11} as used

v. $\mathbf{g}(x) = (x^3 + \alpha^{12}x^2 + \alpha^8 x + \alpha^4)(x + \alpha^{11})$

$\qquad = x^4 + x^3 + 1$

Note: At this point, $\mathbf{g}(x)$ contains the minimal polynomial for elements $\alpha^7, \alpha^{14}, \alpha^{13}$, and α^{11}; the coefficients are elements of GF(2).

5. $i = 3$

A. $k = 6$

B. α^6 has not been used

C. $r = 0$

    i. $\lambda = 6$

    ii. $\lambda = 6$

    iii. continue

    iv. mark α^6 as used

    v. $\mathbf{g}(x) = (x^4 + x^3 + 1)(x + \alpha^6)$

$\qquad = x^5 + \alpha^{13}x^4 + \alpha^6 x^3 + x + \alpha^6$

C. $r = 1$

    i. $\lambda = 12$

    ii. $\lambda = 12$

    iii. continue

    iv. mark α^{12} as used

    v. $\mathbf{g}(x) = (x^5 + \alpha^{13}x^4 + \alpha^6 x^3 + x + \alpha^6)(x + \alpha^{12})$

$\qquad = x^6 + \alpha x^5 + \alpha^7 x^4 + \alpha^3 x^3 + x^2 + \alpha^4 x + \alpha^3$

C. $r = 2$

    i. $\lambda = 24$

    ii. $\lambda = 9$

    iii. continue

    iv. mark α^9 as used

    v. $\mathbf{g}(x) = (x^6 + \alpha x^5 + \alpha^7 x^4 + \alpha^3 x^3 + x^2 + \alpha^4 x + \alpha^3)(x + \alpha^9)$

$\qquad = x^7 + \alpha^3 x^6 + \alpha^6 x^5 + \alpha^9 x^4 + \alpha^{11}x^3 + \alpha^{14}x^2$

$\qquad\qquad + \alpha^8 x + \alpha^{12}$

C. $r = 3$

    i. $\lambda = 48$

    ii. $\lambda = 3$

    iii. continue

    iv. mark α^3 as used

    v. $\mathbf{g}(x) = (x^7 + \alpha^3 x^6 + \alpha^6 x^5 + \alpha^9 x^4 + \alpha^{11}x^3 + \alpha^{14}x^2$

$\qquad\qquad + \alpha^8 x + \alpha^{12})(x + \alpha^3)$

$\qquad = x^8 + x^4 + x^2 + x + 1$

Note: At this point, $\mathbf{g}(x)$ contains the product of the minimal polynomial for α^7, α^{11}, α^{13}, α^{14} and the minimal polynomial for α^3, α^6, α^9, α^{12}.

5. $i = 5$
   A. $k = 5$
   B. α^5 has not been used
   C. $r = 0$
      i. $\lambda = 5$
      ii. $\lambda = 5$
      iii. continue
      iv. mark α^5 as used
      v. $\mathbf{g}(x) = (x^8 + x^4 + x^2 + x + 1)(x + \alpha^5)$
      $= x^9 + \alpha^5 x^8 + x^5 + \alpha^5 x^4 + x^3 + \alpha^{10}x^2 + \alpha^{10}x + \alpha^5$
   C. $r = 1$
      i. $\lambda = 10$
      ii. $\lambda = 10$
      iii. continue
      iv. mark α^{10} as used
      v. $\mathbf{g}(x) = (x^9 + \alpha^5 x^8 + x^5 + \alpha^5 x^4 + x^3 + \alpha^{10}x^2 + \alpha^{10}x +$
      $\alpha^5)(x + \alpha^{10})$
      $= x^{10} + x^9 + x^8 + x^6 + x^5 + x^2 + 1$
   C. $r = 2$
      i. $\lambda = 20$
      ii. $\lambda = 5$
      iii. $\lambda = k$ and $r \neq 0$, so we set i to the next odd value and
           continue at step 5A. The next value of i would be 7,
           which is greater than $2\tau - 1$; therefore, step 5 has now
           been completed.
6. The result $\mathbf{g}(x) = x^{10} + x^9 + x^8 + x^6 + x^5 + x^2 + 1$ is the desired
   generator polynomial. This result matches the result of
   Example 4.1.

Listing 4-1

```
/****************************/
/*                          */
/*    BuildBchCode()        */
/*                          */
/****************************/

void BuildBchCode(   int first_reqd_alpha_power,
                     int last_reqd_alpha_power,
                     int alpha_power_increm)

{
int i,j;
logical include_power[64];
word required_min_poly, working_gen_poly;

for( i=first_reqd_alpha_power; i<=last_reqd_alpha_power;
     i+=alpha_power_increm) include_power[i]=FALSE;
Degree_Of_Gen_Poly = 0;
```

```
working_gen_poly = 1;
for( i=first_reqd_alpha_power; i<=last_reqd_alpha_power;
             i+=alpha_power_increm) {
  if(include_power[i]==FALSE) {
  printf("include_power[%d]==FALSE\n",i);
    /*------------------------------------------------------------*/
    /*  Consider in sequence each power of alpha that is          */
    /*  to be a root of the generator polonyomial.  If a          */
    /*  particular power of alpha is not included yet,            */
    /*  multiply the generator polynomial under construction      */
    /*  by the minimal polynomial corresponding to this power     */
    /*  of alpha.  Mark this power and all the other roots of     */
    /*  this minimal polynomial as now being included among       */
    /*  the roots of the generator polynomial.                    */
    required_min_poly = GetMinPoly(i);
    Degree_Of_Gen_Poly += GetDegreeOfMinPoly(i);
    printf("required_min_poly set to %d\n",required_min_poly);
    include_power[i]=TRUE;
    for( j=i+1; j<=last_reqd_alpha_power; j++) {
      if(GetMinPoly(j)==required_min_poly) {
        include_power[j]=TRUE;
      }
    }
    ShiftRegPolyMultipler(   &working_gen_poly,
                             required_min_poly);
  }
}
Gen_Poly_High_Left = working_gen_poly;
Gen_Poly_High_Right = FlipPackedPoly(working_gen_poly);

printf("Gen_Poly_High_Left = %#lx\n",Gen_Poly_High_Left);
printf("Gen_Poly_High_Right = %#lx\n",Gen_Poly_High_Right);
printf("Degree_Of_Gen_Poly = %d\n",Degree_Of_Gen_Poly);
return;
}
```

4.2 Nonbinary BCH Codes

It is possible to extend the concept of a BCH code to include nonbinary codes. As covered in Section 4.1, the generator polynomial $g(x)$ for a t-error-correcting binary BCH code is intentionally architected to have among its roots $2t$ consecutive powers of β where β is some element of the code's locator field $GF(2^m)$. However, to ensure that the coefficients of $\mathbf{g}(x)$ are elements of the symbol field $\mathbf{GF}(2)$, additional roots might need to be included. Thus the generator polynomial can be expressed as

$$\mathbf{g}(x) = \text{LCM}\{M^{(i)}(x), M^{(i+1)}(x), ..., M^{(i+2t-1)}(x)\}$$

where $M^{(i)}(x)$ is the minimal polynomial of element β^i. When β is a primitive element (*i.e.*, the order of β is $2^m - 1$) the code is a primitive BCH code, and when β is not a primitive element (*i.e.*, the order of β is less than $2^m - 1$) the code is a nonprimitive BCH code. In either case, the code's length n equals the order of β.

Similar relationships can be established for nonbinary BCH codes. For a t-error-correcting nonbinary BCH code with locator field GF(q^s) and symbol field GF(q), the generator polynomial is architected to have among its roots $2t$ consecutive powers of β where β is some element of the locator field GF(q^s). If β is a primitive element of GF(q^s), its order is $q^s - 1$, and the code's length is $n = q^s - 1$ q-ary symbols. Each symbol is an element of the symbol field GF(q). If β is a nonprimitive element of GF(q^s), its order is a divisor of $q^s - 1$, and the code's length n will equal the order of β.

Example 4.5 Build a generator polynomial for a primitive nonbinary $(8, k, t = 1)$ BCH code over GF(3^2).

Solution The extension field GF(3^2) generated by $\alpha^2 = 2\alpha + 1$ is listed in Table 4-3. The primitive elements of this field are α, α^3, α^5, and α^7. The elements α^2 and α^6 are of order 4, and the element α^4 is of order 2. Addition on this field is defined in Table 4-4. For $\beta = \alpha$, we have

$$\mathbf{g}(x) = \text{LCM}\{M^{(1)}(x), M^{(2)}(x)\} \tag{4.5}$$

■ **Table 4-3** Extension field GF(3^2) generated by $\alpha^2 = 2\alpha + 1$.

Power form	2-tuple form
$-\infty$	0 0
α^0	0 1
α^1	1 0
α^2	2 1
α^3	2 2
α^4	0 2
α^5	2 0
α^6	1 2
α^7	1 1

		$-\infty$	α^0	α^1	α^2	α^3	α^4	α^5	α^6	α^7
		00	01	10	21	22	02	20	12	11
$-\infty$	00	00	01	10	21	22	02	20	12	11
α^0	01	01	02	11	22	20	00	21	10	12
α^1	10	10	11	20	01	02	12	00	22	21
α^2	21	21	22	01	12	10	20	11	00	02
α^3	22	22	20	02	10	11	21	12	01	00
α^4	02	02	00	12	20	21	01	22	11	10
α^5	20	20	21	00	11	12	22	10	02	01
α^6	12	12	10	22	00	01	11	02	21	20
α^7	11	11	12	21	02	00	10	01	20	22

The order of β is $3^2 - 1$, which divides into cyclotomic cosets as

$$C_0 = \{0\} \qquad C_4 = \{4\}$$
$$C_1 = \{1, 3\} \qquad C_5 = \{5, 7\}$$
$$C_2 = \{2, 6\}$$

Therefore, Eq. 4.5 can be simplified to

$$\mathbf{g}(x) = M^{(1)}(x)M^{(2)}(x) \qquad (4.6)$$

where

$$M^{(1)}(x) = (x - \alpha^1)(x - \alpha^3) \qquad (4.7)$$
$$= x^2 + (\alpha^5 + \alpha^7)x + \alpha^4$$
$$= x^2 + x + 2$$

and

$$M^{(2)}(x) = (x - \alpha^2)(x - \alpha^6) \qquad (4.8)$$
$$= x^2 + (\alpha^6 + \alpha^2)x + \alpha^8$$
$$= x^2 + 1$$

Substituting Eqs. 4.7 and 4.8 into 4.6, we obtain

$$\mathbf{g}(x) = (x^2 + x + 2)(x^2 + 1)$$
$$= x^4 + x^3 + x + 2$$

Because the degree of $\mathbf{g}(x)$ is 4, we obtain k as $k = n - 4 = 8 - 4 = 4$.

Example 4.6 Build a generator polynomial for a nonprimitive nonbinary $(4, k, t = 1)$ BCH code over GF(3^2).

Solution A nonprimitive element of order 4 from GF(3^2) is $\beta = \alpha^2$. Therefore, we obtain $\mathbf{g}(x)$ as

$$\mathbf{g}(x) = \text{LCM}\{M^{(2)}(x), M^{(4)}(x)\}$$

where

$$M^{(2)}(x) = x^2 + 1$$
$$M^{(4)}(x) = x + 1$$

so

$$\mathbf{g}(x) = (x^2 + 1)(x + 1)$$
$$= x^3 + x^2 + x + 1$$

4.3 Reed-Solomon Codes

As discussed in Section 4.2, associated with every BCH code is a locator field $\mathrm{GF}(q^s)$ and a symbol field $\mathrm{GF}(q)$, where $q = p^m$, with p prime. Binary BCH codes are simply a special case obtained when $q = 2$. Another subset of BCH codes, called *Reed-Solomon codes*, is obtained when $s = 1$, and the locator field and symbol field are the same $\mathrm{GF}(q)$.

In previous sections, it was necessary to form cyclotomic cosets and generate minimal polynomials having roots in $\mathrm{GF}(q^s)$ and coefficients in $\mathrm{GF}(q)$. A consequence of setting $s = 1$ is that all minimal polynomials are linear; in other words, a binomial such as $(x - \alpha^i)$ is already a minimal polynomial for all α^i in $\mathrm{GF}(q)$. Thus, the generator polynomial for a t-error-correcting [or design-distance $(2t + 1)$] Reed-Solomon code can be expressed as

$$\mathbf{g}(x) = (x - \alpha^i)(x - \alpha^{i+1})...(x - \alpha^{i+2t-1})$$
$$= \prod_{\lambda = i}^{i+2t-1} (x - \alpha^\lambda)$$
$$= \prod_{\lambda = 0}^{2t-1} (x - \alpha^{i+\lambda})$$

A point not made clear in some coding books is that we still have $q = p^m$, and m can be set to 1 to obtain codes over prime fields, such as $\mathrm{GF}(11)$ and $\mathrm{GF}(17)$. In many cases, Reed-Solomon codes, unless indicated otherwise, are assumed to have $\mathrm{GF}(2^m)$ as both their locator field and symbol field.

Example 4.7 Build a generator polynomial for a nonprimitive nonbinary $(7, 2, d = 6)$ BCH code over $\mathrm{GF}(8)$.

Solution Addition and multiplication over $\mathrm{GF}(8)$ are defined in Tables 4-5 and 4-6. Select the primitive element $\beta = a$ and obtain the generator polynomial as

$$\mathbf{g}(x) = \prod_{\lambda = 1}^{5} (x - \beta^\lambda) \tag{4.9}$$
$$= (x - a)(x - b)(x - c)(x - d)(x - e)$$

+	0	1	a	b	c	d	e	f
0	0	1	a	b	c	d	e	f
1	1	0	c	f	a	e	d	b
a	a	c	0	d	1	b	f	e
b	b	f	d	0	e	a	c	1
c	c	a	1	e	0	f	b	d
d	d	e	b	a	f	0	1	c
e	e	d	f	c	b	1	0	a
f	f	b	e	1	d	c	a	0

■ Table 4-6 Multiplication over GF(8).

×	0	1	a	b	c	d	e	f
0	0	0	0	0	0	0	0	0
1	0	1	a	b	c	d	e	f
a	0	a	b	c	d	e	f	1
b	0	b	c	d	e	f	1	a
c	0	c	d	e	f	1	a	b
d	0	d	e	f	1	a	b	c
e	0	e	f	1	a	b	c	d
f	0	f	1	a	b	c	d	e

83

Because Table 4-5 shows each element to be its own additive inverse, we can rewrite Eq. 4.9 as

$$\mathbf{g}(x) = (x + a)(x + b)(x + c)(x + d)(x + e)$$
$$= x^5 + bx^4 + cx^3 + fx^2 + dx + a$$

The resulting code is shown in Table 4-7.

Info digits	Check digits	Info digits	Check digits
0 0	0 0 0 0 0	c 0	b d e a f
0 1	b c f d a	c 1	0 f a b e
0 a	c d 1 e b	c a	e 0 d f 1
0 b	d e a f c	c b	a 1 f e d
0 c	e f b 1 d	c c	c c c c c
0 d	f 1 c a e	c d	1 e b 0 a
0 e	1 a d b f	c e	f b 1 d 0
0 f	a b e c 1	c f	d a 0 1 b
1 0	f a b e c	d 0	c e f b 1
1 1	1 1 1 1 1	d 1	e b 0 a c
1 a	d b f 0 e	d a	0 1 b c f
1 b	c f d a 0	d b	f 0 e 1 a
1 c	a e 0 d f	d c	b a 1 f e
1 d	0 c e f b	d d	d d d d d
1 e	b 0 a c d	d e	a f c 0 b
1 f	e d c b a	d f	1 c a e 0
a 0	1 b c f d	e 0	d f 1 c a
a 1	f e d c b	e 1	a d b f 0
a a	a a a a a	e a	f c 0 b d
a b	e c 1 0 f	e b	0 a c d 1
a c	d 1 e b 0	e c	1 0 f a b
a d	b f 0 e 1	e d	c b a 1 f
a e	0 d f 1 c	e e	e e e e e
a f	c 0 b d e	e f	b 1 d 0 c
b 0	a c d 1 e	f 0	e 1 a d b
b 1	d 0 c e f	f 1	c a e 0 d
b a	1 f e d c	f a	b e c 1 0
b b	b b b b b	f b	1 d 0 c e
b c	f d a 0 1	f c	0 b d e a
b d	e a f c 0	f d	a 0 1 b c
b e	c 1 0 f a	f e	d c b a 1
b f	0 e 1 a d	f f	f f f f f

84

Encoders and Decoders

5

As discussed in Chapter 3, the encoding operation for block codes can be cast as a matrix multiplication involving the generator matrix $\mathbf{G}$

$$\mathbf{v} = \mathbf{uG}$$

For any given information field, we can manually generate the corresponding parity field using Recipe 3.1 from Section 3.4. However, for practical applications, we must find methods that automate the encoding and decoding operations. Such methods are the subject of this chapter.

5.1 Division Method for Encoding Cyclic Codes

Section 3.4 presents Recipe 3.1, which can be used to manually encode cyclic codes. In terms of polynomial representations, this Recipe takes the information polynomial $\mathbf{u}(x)$, multiplies it by x^{n-k}, and finally divides the result $x^{n-k}\mathbf{u}(x)$ by the generator polynomial $\mathbf{g}(x)$. The remainder $\mathbf{r}(x)$ of this division operation is the polynomial representation of the $(n-k)$-bit check field.

5.1.1 Hardware Implementation

A sequential logic circuit that accomplishes this same set of operations is shown in Fig. 5-1. Each of the square boxes labeled with a power of x represents one stage of a shift register. The total number of stages is equal to the degree of the generator polynomial and the number of check bits in each codeword. Each of the circled coefficients $g_0, g_1, \cdots g_{n-k-1}$ represents a binary multiplier. If the generator polynomial contains a term for x^i, the coefficient g_i equals 1. If the term for x^i is missing from the generator polynomial, the coefficient g_i equals 0. The parity tree outputs a 1 if and only if an odd number of its inputs are set to 1. The input sequence consists of the k information bits followed by $n-k$ zeros. On each cycle of operation, one of the k information bits is clocked into the input, with the enable line turned on. The output of the parity tree

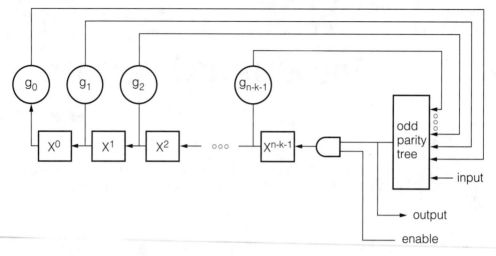

■ **5-1** *Circuit for encoding of cyclic codes.*

will be issued as an output bit and also shifted into the rightmost stage of the shift register. After k cycles, the enable line is turned off so that on subsequent cycles, zeros are shifted into the rightmost stage of the shift register. The first k output bits will represent coefficients of the quotient polynomial $\mathbf{q}(x)$, where

$$\mathbf{u}(x)x^{n-k} = \mathbf{g}(x)q(x) + \mathbf{r}(x)$$

The $n - k$ output bits produced after the enable line is turned off will represent coefficients of the remainder polynomial $\mathbf{r}(x)$. These coefficients of $\mathbf{r}(x)$ are the check bits for the codeword. Operation of this circuit is best illustrated by means of several examples.

Example 5.1 Draw a circuit like Fig. 5-1 tailored to the specific case of an (n, k) code with generator polynomial $\mathbf{g}(x) = x^3 + x^2 + 1$. Construct a table showing the input bit, enable status, shift register contents, and output bit at each cycle of operation for the information sequence 0100.

Solution The coefficients g_i will each have a value of 0 or 1, so the actual implementation of Fig. 5-1 for a specific $\mathbf{g}(x)$ will simply make a connection between the shift register and the parity tree for those stages having their $g_i = 1$. The connection is omitted for each stage having $g_i = 0$. The desired circuit is shown in Fig. 5-2. The input, output, enable status, and shift register contents for each cycle are listed in Table 5-1. The last 3 output bits constitute the check field which is 011.

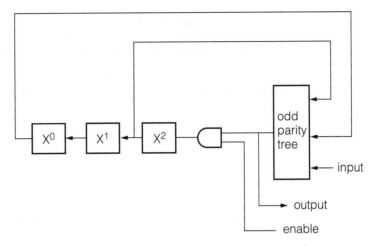

■ **5-2** *Circuit for encoding the (n, k) cyclic code having* **g**(x)=x³ + x² +1.

■ Table 5-1 Details of
encoder operation for Example 5.1.

Cycle	Input	Enable status	Shift register contents	Output
1	0	on	0 0 0	0
2	1	on	0 0 1	1
3	0	on	0 1 1	1
4	0	on	1 1 1	1
5	0	off	1 1 0	0
6	0	off	1 0 0	1
7	0	off	0 0 0	1

Example 5.2 Use the circuit of Fig. 5-2, and construct a table showing the input bit, enable status, shift register contents, and output bit at each cycle of operation for the information sequence 0011.

Solution The desired information is listed in Table 5-2. The last 3 output bits constitute the check field which is 010.

Cycle	Input	Enable status	Shift register contents	Output
1	0	on	0 0 0	0
2	0	on	0 0 0	0
3	1	on	0 0 1	1
4	1	on	0 1 0	0
5	0	off	1 0 0	0
6	0	off	0 0 0	1
7	0	off	0 0 0	0

5.1.2 Software Implementation

A function cyclicEncoder(), that emulates the operation of the sequential circuit from Fig. 5-1 is provided in Listing 5-1.

Listing 5-1

```
/************************/
/*                      */
/*   CyclicEncoder      */
/*                      */
/************************/

void CyclicEncoder( word gen_poly,
                    int degree_of_gen_poly,
                    bits code_vector[],
                    int num_info_bits)
{
int n,bit_pointer;
unsigned long int shift_register, output_register, tap_mask;
unsigned long int active_register_mask;
unsigned short int parity_bit,new_bit;
/*----------------------------------------------------------*/
/*  Pack the generator polynomial's individual coeffcients  */
/*  into a single unsigned long int word.  Result is right- */
/*  justified with coefficient for the term whose degree is */
/*  (degree_of_gen_poly - 1) in the least significant bit.   */

tap_mask = gen_poly >>1;
active_register_mask = (unsigned long int)
                (IntPower(2,degree_of_gen_poly)-1);
/*----------------------------------------------------------*/
/*  shift register operation with gate enabled              */
/*  This mode is used for num_info_bits clock cycles         */
/*  In addition to being input to the shift register,       */
/*  the input information bits are copied from the           */
/*  low-index locations of code_vector[] into the            */
/*  appropriate high-index locations so that the            */
/*  degree_of_gen_poly lowest-index locations will be       */
```

```
/*  available for the check bits that will be flushed  */
/*  from the shift register after the gate is disabled. */

bit_pointer = degree_of_gen_poly + num_info_bits;
shift_register = 0;
output_register = 0;
for(n=num_info_bits;n>0;n--) {
  parity_bit = LongParity((tap_mask&shift_register));
  new_bit = parity_bit ^ code_vector[n];
  bit_pointer--;
  code_vector[bit_pointer] = code_vector[n];
  shift_register = shift_register << 1;
  shift_register = shift_register ^ (long)new_bit;
  output_register = output_register << 1;
  output_register = output_register ^ (long)new_bit;
  /* at this point, output_register contains the desired
quotient */
  /* this quotient is q(x) for which

        b(x)*(x**m) = g(x)*q(x) + r(x)
      where
        g(x) = generator polynomial
        m = degree of generator polynomial
        r(x) = remainder polynomial
        b(x) = polynomial representation of info bits
  */
  }

/*-------------------------------------------------*/
/*  shift register operation with gate disabled  */
output_register = 0;
for( n=degree_of_gen_poly; n>0; n--) {
  parity_bit = LongParity((tap_mask&shift_register));
  bit_pointer--;
  code_vector[bit_pointer] = parity_bit;
  shift_register = shift_register << 1;
  output_register = output_register << 1;
  output_register = output_register ^ (long)parity_bit;
  }
return;
}
```

5.2 Standard Array

For every block code, there is a particular matrix called the *standard array*. For small codes, decoding can be accomplished via direct use of the standard array. However, for large codes, direct use of the standard array becomes prohibitively cumbersome. Nevertheless, the basic concepts embodied in the standard array are important because they form the theoretical basis for more computationally efficient decoding algorithms. For a code with k information bits and n total bits, the standard array will contain 2^k columns and 2^{n-k} rows.

Recipe 5.1 Constructing the standard array.

1. Using one code word per column, construct the first row of the standard array. Place the all-zero code word in the first location of this row.

2. Pick an m-tuple of minimum weight that is not already included in the standard array. This will become the first element of a new row.

3. Complete the new row with elements formed by adding [bit-by-bit over GF(2)] the first element to the corresponding element of the first row. (See Fig. 5-2.)

4. Repeat steps 2 and 3 until the standard array contains all 2^m possible m-tuples.

Definition 5.1 Each row in the standard array is called a *coset*.

Definition 5.2 The first element in each row of the standard array is called the *coset leader* for that row.

Example 5.3 Use Recipe 5.1 to construct a standard array for the (7, 3, d = 4) BCH code having $\mathbf{g}(x) = x^4 + x^3 + x^2 + 1$.

Solution

1. Using $\mathbf{g}(x)$ along with any of the manual encoding techniques from Section 3.4, we can obtain the eight 7-bit code words for the desired code as 0000000, 0011101, 0100111, 0111010, 1001110, 1010011, 1101001, and 1110100. (These code words are in systematic form with the leftmost 3 bits being the information field and the rightmost 4 bits being the parity field.)

2. In accordance with step 1 of Recipe 5.1, the first row of the standard array contains 8 columns, with one codeword in each column (See Table 5-3).

3. Perform steps 2 and 3 of the Recipe seven times—once for each of the seven possible 7-tuples of weight 1. This will generate rows 2 through 8 as shown in Table 5-3.

4. The complete standard array must contain $2^7 = 128$ entries, but so far we have exhausted all the 7-tuples of weight 1 and generated only 64 entries. We now must start using 7-tuples of weight 2 as coset leaders. There are 21 different 7-tuples of weight 2, and not all of them will be used as coset leaders. One arbitrary selection results in rows 9 through 15, as shown in Table 5-3.

■ Table 5-3 Standard array for the (7, 3, $d = 4$) BCH code defined by $g(x) = x^4 + x^3 + x^2 + 1$.

5. By the time we complete row 15, all 7-tuples of weight 2 have been placed in the standard array. Therefore, for the coset leader of row 16, we must select one of the 7-tuples of weight 3 that has not appeared elsewhere in the standard array.

Decoding with the standard array is accomplished by using Recipe 5.2.

Recipe 5.2 Decoding with the standard array.

1. Find the m-tuple within the standard array that matches the received m-tuple. When the match is located in row i column j of the standard array, denote the matching m-tuple as $M_{i,j}$.
2. Use the first m-tuple in column j as the decoded output. This output is denoted as $M_{1,j}$. For this decoded output, the assumed error pattern is given by $M_{i,1}$ (the first m-tuple in row i).

Example 5.4 Use the standard array given in Table 5-3 to decode the received m-tuples 0110010 and 1101001.

Solution Searching the table, we find that the m-tuple 0110010 is located in row 5 column 4. The coset leader for row 5 is 0001000, indicating that an error has occurred in the middle bit position. The decoded output is given by the m-tuple in row 1 column 4, which is 0111010. The second received m-tuple is located in row 1 column 7 of the standard array. The coset leader for row 1 is 0000000 indicating that any received m-tuple found in row 1 is deemed to be error-free; the decoded output equals the received m-tuple.

5.2.1 Hazards in Standard Array Decoding

When some $(t + 1)$ bit errors are correctable, it might make a difference to some applications just how this capability is utilized. One common approach is to select the coset leaders with weight $(t + 1)$ such that the correctable errors lie within the information bits. The usually stated reasoning is that this is done to provide "more protection" to the information bits. Unfortunately, there is a fallacy in the reasoning behind such an approach. Two situations can occur:

1. $t + 1$ errors will occur, aligned in one of the correctable patterns.
   - The received word will match one of the entries in the lower portion of the standard array (i.e., the portion where the coset leaders have weight $t + 1$).

- The decoder will correctly assume that the errors occur in the pattern indicated by the applicable coset leader and will make corrections accordingly.

2. $t + 1$ errors will occur, aligned in an uncorrectable pattern.
   - The received word will match one of the entries in the lower portion of the standard array.
   - The decoder will incorrectly assume that the errors occur in the particular pattern indicated by the applicable coset leader and will make "corrections" accordingly. In general, the attempted "corrections" will, in fact, not be corrections at all, and in many cases the decoder will increase the number of errors in the information field.

Example 5.5 Assume that the m-tuple 0011101 is transmitted and that errors occur in the second and fifth bit positions from the left so that the received m-tuple is 0111001. Show how standard-array decoding using Table 5-3 results in an incorrect decoding.

Solution The received m-tuple 0111001 is found in row 13 column 7 of Table 5-3. The corresponding coset leader is 1010000 indicating that errors have occurred in the first and third bits from the left. In accordance with Recipe 5.2, this results in a decoded output of 1101001. Comparison of this result with the original transmitted m-tuple reveals that the decoded output contains errors in the first, second, third, and fifth bit positions. By attempting to fix two errors, we have wound up with four! The situation is even worse if we only consider the impact on the information bits. The original error corrupts one information bit and one check bit. After "correction" the m-tuple contains 3 incorrect information bits and one incorrect check bit.

As shown by the example, two-bit errors are not all **correctable** with this code. However, they are all **detectable**. We can draw a horizontal line in Table 5-3 between the last coset leader with weight 1 and the first coset leader with weight 2. We can draw a second line between the last coset leader with weight 2 and the first coset leader with weight 3. Any received words falling between these two lines are assumed to have two bit errors in an unknown pattern, which might or might not be correctable.

5.3 Syndromes for BCH Codes

By definition, the generator polynomial of a BCH code has $2t$ consecutive powers of a as roots, where α is a primitive element in the extension field $GF(2^m)$. Each codeword polynomial $c(x)$ will have the same sequence of roots, or in other words

$$c(\alpha^i) = 0 \quad i = 1, 2, 3, ..., 2t - 1 \tag{5.1}$$

where

$$c(x) \equiv c_0 + c_1 x + c_2 x^2 + ... + c_{n-1} x^{n-1}$$

Consequently, we can evaluate $c(x)$ at each odd power of α and thereby construct the following set of simultaneous equations

$$
\begin{aligned}
c_0 \alpha^0 + c_1 \alpha^1 + c_2 \alpha^2 + ... + c_{n-1} \alpha^{n-1} &= 0 \\
c_0 \alpha^0 + c_1 \alpha^3 + c_2 \alpha^6 + ... + c_{n-1} \alpha^{3(n-1)} &= 0 \\
\vdots \qquad\qquad\qquad\qquad\qquad & \quad \vdots \\
c_0 \alpha^0 + c_1 \alpha^{2t-1} + c_2 \alpha^{2(2t-1)} + ... + c_{n-1} \alpha^{(n-1)(2t-1)} &= 0
\end{aligned}
$$

Let's consider a single received word that can be represented as a polynomial $r(x)$, which is the sum of the transmitted codeword polynomial $c(x)$ and the error polynomial $e(x)$

$$r(x) = c(x) + e(x) \tag{5.2}$$

We can evaluate $r(x)$ for each of the roots $\alpha, \alpha^3, \alpha^5, ..., \alpha^{2t-1}$ and obtain

$$r(\alpha^k) = c(\alpha^k) + e(\alpha^k) \quad k = 1, 3, ..., 2t - 1$$

However, we know from Eq. 5.1 that $c(\alpha^k) = 0$ for $k = 1, 3, ..., 2t - 1$, so we are left with

$$r(\alpha^k) = e(\alpha^k) \tag{5.3}$$

The values produced by Eq. 5-3 are called *syndrome* values and are usually denoted as s_k:

$$
\begin{aligned}
s_k &= r(\alpha^k) = e(\alpha^k) \\
&= e_0 (\alpha^k)^0 + e_1 (\alpha^k)^1 + e_2 (\alpha^k)^2 + ... + e_{n-1} (\alpha^k)^{n-1}
\end{aligned}
$$

The coefficients e_i will be either 0 or 1, so the syndrome value can be expressed as a sum of just the terms having nonzero coefficients

$$s_k = \sum_{e_i \neq 0} (\alpha^k)^i \quad k = 1, 3, ..., 2t - 1$$

Now it is a simple matter to reverse the order of the exponents on the summand to obtain

$$s_k = \sum_{e_i \neq 0} (\alpha^i)^k \quad k = 1, 3, ..., 2t - 1 \tag{5.4}$$

Equation 5.4 actually defines a system of equations that, given s_k, could be solved for the nonzero coefficients e_i.

$$s_1 = e_0\alpha^0 + e_1\alpha^1 + e_2\alpha^2 + ... + e_{n-1}\alpha^{n-1} \qquad (5.5)$$
$$s_3 = e_0\alpha^0 + e_1\alpha^3 + e_2\alpha^6 + ... + e_{n-1}\alpha^{3(n-1)}$$
$$s_5 = e_0\alpha^0 + e_1\alpha^5 + e_2\alpha^{10} + ... + e_{n-1}\alpha^{5(n-1)}$$
$$\vdots \qquad \vdots \qquad \vdots$$
$$s_{2t-1} = e_0\alpha^0 + e_1\alpha^{2t-1} + e_2\alpha^{2(2t-1)} + ... + e_{n-1}\alpha^{(n-1)(2t-1)}$$

Unfortunately, this system is somewhat harder to solve than it looks. The hidden difficulties are largely caused by the "modulo" nature of arithmetic over $GF(2^m)$ that is used in computing these equations.

Assume that we have a codeword of n bits and that a gremlin somewhere has the job of inserting t randomly placed errors in this block of n bits. For locating the first error, the gremlin can choose from any of the n possible bit positions. For locating the second error, the choices are limited to the $n-1$ positions not containing the first error. For the third error, the choices are limited to the $n-2$ bit positions not containing either of the first two errors. The total number of error-placement **sequences** for placing t errors in a word of n bits is thus given by

$$(n)(n-1)(n-2) \, ... \, (n-t+1) = \frac{n!}{(n-t)!}$$

For any given t-bit error **pattern**, any one of the t bits could have been the first one placed by the gremlin, any one of the remaining $t-1$ could have been second, and so on. The number of ways that the t bits can be ordered is $t!$. However, all we care about is where the errors are located—not the particular order in which the gremlin happened to select these locations. Therefore, to obtain the total number of t-bit error patterns, we must divide the total number of error-placement sequences by the number of ways that t bits can be ordered.

$$N_p = \frac{n!}{(n-t)!t!} \equiv \binom{n}{t} \qquad (5.6)$$

Each of the s_i in Eq. 5.5 will be an element of $GF(2^m)$ where $n = 2^m - 1$, so there are only $n+1$ possible values for s_i. Based on Eq. 5.6, we can see that in most practical situations, the number of error patterns will be larger than the number of possible values for s_i. Therefore, we conclude that several different error patterns can generate the same values of s_i for any particular i. Thus, given the value for a single s_i (computed by substituting a single root of $g(x)$ into the polynomial corresponding to the received word) it will not be possible to solve for the actual error pattern. However, it can be shown that when a set of equations is formed by substituting **each** of the roots of $g(x)$ into $r(x)$, the resulting system of equations can be solved to provide the error pattern. (Assuming of course that the number of error does not exceed *t*.) The Peterson-Berlekamp Method is one of the techniques that can be used to solve a system of syndrome equations.

Software Notes Listing 5-2 contains the ComputeSyndrome() function that computes a set of syndromes in accordance with Eq. 5.5. This function makes use of the PolyDivShiftReg() function to emulate the operation of a shift register circuit for polynomial division.

Listing 5-2

```
/****************************/
/*                          */
/*   ComputeSyndrome        */
/*                          */
/****************************/

void ComputeSyndrome( bits received_poly[],
                      int degree_of_received_poly,
                      exponent syndrome_vector[],
                      exponent first_alpha_expon,
                      exponent last_alpha_expon,
                      exponent alpha_exp_increm)
{
exponent i;
word min_poly_packed;
bits min_poly_vector[MAX_TERMS];
int degree_of_min_poly,first_bit;
logical high_left;
bits quotient_poly[32], remainder_poly[32];

for (i=first_alpha_expon; i<=last_alpha_expon;
i+=alpha_exp_increm) {
  min_poly_packed = GetMinPoly(i);
  degree_of_min_poly = DegOfPackedPoly(min_poly_packed);
  first_bit=0;
  high_left=TRUE;

  UnpackBitsLong( min_poly_packed,
                  min_poly_vector,
                  degree_of_min_poly+1,
                  first_bit,
                  high_left);

  PolyDivShiftReg( min_poly_vector,
                   degree_of_min_poly,
                   received_poly,
                   degree_of_received_poly,
                   quotient_poly,
                   remainder_poly);

  syndrome_vector[i] =
EvalBinaryPolyForExpElem(remainder_poly,i);
  }
return;
}
```

5.4 Peterson-Berlekamp Method

As it was defined earlier, the syndrome **s** is the $(2t)$-tuple $\mathbf{s} = (s_1,$ $s_2, ..., s_{2t})$ where each of the components s_k is given by

$$s_k = \sum_{\lambda = 1}^{\nu} (\alpha^\lambda)^k = \sum_{\lambda = 1}^{\nu} (\beta_\lambda)^k \qquad 1 \le k \le 2t$$

Using the (inverse of the error location number) β_k, we define the error location polynomial $\sigma(x)$ as

$$\sigma(x) = (1 + \beta_1 x)(1 + \beta_2 x) \, ... \, (1 + \beta_\nu x) \qquad (5.7)$$

We can perform all the multiplications indicated in Eq. 5.7 to obtain $\sigma(x)$ in unfactored form as

$$\sigma(x) = \sigma_0 + \sigma_1 x + \sigma_2 x^2 + ... + \sigma_\nu x^\nu$$

The Peterson-Berlekamp method is a way to find the error-locator polynomial from the syndrome of a received BCH code word. Hand calculations using this method are made easier and less confusing if a table-filling approach is used.

Recipe 5.3 *Peterson-Berlekamp Algorithm. (Table-filling approach for hand calculation.)*

1. Set up a table with initial conditions as follows:

μ	$\sigma^{(\mu)}(x)$	d_μ	l_μ	$\mu - l_\mu$
-1	1	1	0	-1
0	1	s_1	0	0

   where s_1 is the first component of the syndrome $(2t)$-tuple **s**. The remainder of the algorithm consists of the iterative steps used to fill out the rest of the table.

2. Let k denote the value of μ from the last filled row in the table. Examine d_k and:
   A. If $d_k = 0$ then set

   $$\sigma^{(k+1)}(x) = \sigma^{(k)}(x)$$

   $$l_{k+1} = l_k$$

   B. If $d_k \ne 0$ examine $\mu - l_\mu$ for all the rows prior to row k for which $d_\mu \ne 0$. Let ρ denote the value of μ for the row found to have the largest value of $\mu - l_\mu$ and having $d_\mu \ne 0$. Set:

   $$\sigma^{(k+1)}(x) = \sigma^{(k)}(x) - d_k d_\rho^{-1} x^{k - \rho} \sigma^{(\rho)}(x)$$

   $$l_{k+1} = \max(l_k, l_\rho + k - \rho)$$

3. Set:

$$d_{k+1} = s_{k+2} + \sigma_1^{(k+1)} s_{k+1} + ... + \sigma_{l_{k+1}}^{(k+1)} s_{k+2 - l_{k+1}}$$

4. Repeat steps 2 and 3 until $\sigma^{(2t)}(x)$ is generated. If $\sigma^{(2t)}(x)$ has degree greater than t, there were more than t errors in the received polynomial.

Example 5.6 Use Recipe 5.3 to find the error locator polynomial for $s_1 = s_2 = s_4 = 1$, $s_3 = s_5 = \alpha^{10}$, and $s_6 = \alpha^5$.

Solution

1. Set up the table of initial conditions:

μ	$\sigma^{(\mu)}(x)$	d_μ	l_μ	$\mu - l_\mu$
-1	1	1	0	-1
0	1	1	0	0

For $n = 2^m - 1 = 15$, we find $m = 4$.

2. Build row 1:

$d_0 \neq 0$ so pick $\rho = -1$:

$$\sigma^{(1)}(x) = \sigma^{(0)}(x) - d_0 d_{-1}^{-1} x^1 \sigma^{(-1)}(x)$$
$$= 1 - x$$
$$l_1 = \max\,(0, 1) = 1$$

$$d_1 = s_2 + \sigma_1^{(1)} s_1$$
$$= 1 + 1 = 0$$

Revised table:

μ	$\sigma^{(\mu)}(x)$	d_μ	l_μ	$\mu - l_\mu$
-1	1	1	0	-1
0	1	1	0	0
1	$1 + x$	0	1	0

3. Build row 2: $d_1 = 0$, so

$$\sigma^{(2)}(x) = \sigma^{(1)}(x) = 1 + x$$
$$l_2 = l_1 = 1$$
$$d_2 = s_3 + \sigma_1^{(2)} s_2$$
$$= \alpha^{10} + 1 = \alpha^5$$

Revised table:

μ	$\sigma^{(\mu)}(x)$	d_μ	l_μ	$\mu - l_\mu$
-1	1	1	0	-1
0	1	1	0	0
1	$1 + x$	0	1	0
2	$1 + x$	α^5	1	1

4. Build row 3: $d_2 \neq 0$ so pick $\rho = 0$

$$\sigma^{(3)}(x) = \sigma^{(2)}(x) - d_2 d_0^{-1} x^2 \sigma^{(0)}(x)$$
$$= 1 + x - \alpha^5 x^2$$
$$= 1 + x + \alpha^5 x^2$$

$$l_3 = \max(l_2, l_0 + 2 - 0)$$
$$= \max(1, 2) = 2$$

$$d_3 = s_4 + \sigma_1^{(3)}(x)s_3 + \sigma_2^{(3)}(x)s_2$$
$$= 1 + \alpha^{10} + \alpha^5 = 0$$

Revised table:

μ	$\sigma^{(\mu)}(x)$	d_μ	l_μ	$\mu - l_\mu$
-1	1	1	0	-1
0	1	1	0	0
1	$1 + x$	0	1	0
2	$1 + x$	α^5	1	1
3	$1 + x + \alpha^5 x^2$	0	2	1

5. Build row 4: $d_3 = 0$ so

$$\sigma^{(4)}(x) = \sigma^{(3)}(x) = 1 + x + \alpha^5 x^2$$
$$l_4 = l_3 = 2$$
$$d_4 = s_5 + \sigma_1^{(4)}s_4 + \sigma_2^{(4)}s_3$$
$$= \alpha^{10} + 1 + \alpha^5 \alpha^{10}$$
$$= 1$$

Revised table:

μ	$\sigma^{(\mu)}(x)$	d_μ	l_μ	$\mu - l_\mu$
-1	1	1	0	-1
0	1	1	0	0
1	$1 + x$	0	1	0
2	$1 + x$	α^5	1	1
3	$1 + x + \alpha^5 x^2$	0	2	1
4	$1 + x + \alpha^5 x^2$	α^{10}	2	2

6. Build row 5: $d_4 \neq 0$ so pick $\rho = 2$

$$\sigma^{(5)}(x) = \sigma^{(4)}(x) - (\alpha^{10}/\alpha^5), x^2 \, \sigma^{(2)}(x)$$
$$= 1 + x + \alpha^5 x^3$$

$$l_5 = \max(l_4, l_2 + 4 - 2)$$
$$= \max(2, 3) = 3$$

99

$$d_5 = s_6 + s_5 + \alpha^5 s_3$$
$$= \alpha^5 + \alpha^{10} + \alpha^{15}$$
$$= 0$$

Revised table:

μ	$\sigma^{(\mu)}(x)$	d_μ	l_μ	$\mu - l_\mu$
-1	1	1	0	-1
0	1	1	0	0
1	$1 + x$	0	1	0
2	$1 + x$	α^5	1	1
3	$1 + x + \alpha^5 x^2$	0	2	1
4	$1 + x + \alpha^5 x^2$	α^{10}	2	2
5	$1 + x + \alpha^5 x^3$	0	3	2

The precise algorithm embodied in Recipe 5.3 can be recast in any number of slightly different forms. One common variation subscripts the components of the syndrome **s** starting at 0 rather than 1. Another common variation computes a pointer to the proper row to use when $d_k \neq 0$, rather than conducting a search for the row with $d_\rho \neq 0$ and having the largest value for $\mu - l_\mu$.

Listing 5-3 contains the PerformBerlekamp() function that implements Recipe 5.3.

Listing 5-3

```
/****************************/
/*                          */
/*  PerformBerlekamp        */
/*                          */
/****************************/
void PerformBerlekamp(  exponent syndrome_vector[],
                        int first_reqd_alpha_power,
                        int last_reqd_alpha_power,
                        int alpha_power_increm,
                        exponent sigma[MAX_TERMS])
{
exponent working_sigma[MAX_TERMS][MAX_TERMS];
exponent work_poly[MAX_TERMS];
exponent discrepancy[MAX_TERMS];
exponent dk_over_d_rho, work_element;
int degree[MAX_TERMS];
int lambda;
int delta[MAX_TERMS];
int k,biggest_delta,mu;
int rho;
/*-----------------------*/
/*  Initialize table     */

working_sigma[0][0]=0;
working_sigma[1][0]=0;
```

```
for (k=1;k<=MAX_DEGREES;k++) {
  working_sigma[0][k] = -1;
  working_sigma[1][k] = -1;
  }
discrepancy[0]=0;
discrepancy[1]=syndrome_vector[1];
degree[0]=0;
degree[1]=0;
delta[0]=-1;
delta[1]=0;

for(k=first_reqd_alpha_power;k<=(last_reqd_alpha_power);
          k+=alpha_power_increm)
  {
  /*--------------------------------------------------------------*/
  /*  if at row k, the discrepancy is not zero, then we      */
  /*  must examine all the rows prior to row k for which the */
  /*  discrepancy is not zero to find which of these rows    */
  /*  has the largest value of delta=mu-degree.  The index   */
  /*  of this row will be designated as rho.                 */

  if( discrepancy[k] >= 0)
    {
    biggest_delta = -2;
    rho=-1;
    for( mu=0; mu<k; mu++)
      {
      if( discrepancy[mu] < 0) continue;
      if( delta[mu] <= biggest_delta) continue;
      biggest_delta=delta[mu];
      rho=mu;
      }
    if(rho<0)
      {
      printf("FATAL ERROR\n");
      return;
      }
    /*------------------------------------------------*/
    /* build the partial error locator polynomial    */
    /*  working_sigma[k+1] as a function of working_sigma[rho] and  */
    /*  working_sigma[k].  Remember: all of these working_sigmas are */
    /*  polynomials in x with coefficients drawn      */
    /*   from GF(2**m).                               */
    CopyExpPolyVec(work_poly,working_sigma[rho]);

    ShiftExpPolyVec(work_poly, k-rho);

    dk_over_d_rho = DivExponFieldElems( discrepancy[k],
                                        discrepancy[rho]);

    MultPolyVecByFieldElem( work_poly, dk_over_d_rho);

    printf("A) working_sigma[%d]:\n",k+1);
    AddExpPolyVecs( working_sigma[k],work_poly,working_sigma[k+1]);
    DumpExpPoly(working_sigma[k+1]);

    /*------------------------------------------*/
    degree[k+1]=degree[rho]+k-rho;
    if(degree[k]>degree[k+1]) degree[k+1]=degree[k];
```

Peterson-Berlekamp Method

```
      }
   /*--------------------------------------------------*/
   /*  if at row k, the discrepancy is zero then   */
   else {
      CopyExpPolyVec(working_sigma[k+1],working_sigma[k]);
      degree[k+1] = degree[k];

   }
   /*-----------------------------------------*/
   /*  compute new delta                   */
   delta[k+1] = k - degree[k+1];

   /*------------------------------------------------*/
   /*  compute the new discrepancy              */
   /*   IAW step 3 of Recipe 8.3.1             */
   discrepancy[k+1]=-1;
   for(lambda=0; lambda<=degree[k+1]; lambda++) {
      work_element = MultExponFieldElems(
syndrome_vector[k+1-lambda],
                     working_sigma[k+1][lambda]);
      discrepancy[k+1]=AddExponFieldElems(discrepancy[k+1],
work_element);
      }
   }
   /*----------------------------------------------------------*/
   /*  copy final row of working matrix into output vector   */
   CopyExpPolyVec(sigma,working_sigma[last_reqd_alpha_power]);
   return;
}
```

102

5.5 Error Location

The output of the Peterson-Berlekamp method is the error location polynomial

$$\sigma(x) = \sigma_0 + \sigma_1 x + \sigma_2 x^2 + ... + \sigma_\nu x^\nu \qquad (5.8)$$

The roots of this polynomial will be the inverse error location numbers. This means that once we find the roots of Eq. 5.8, we only
need to find the inverse of each root in order to determine the location of the erroneous bits. Listing 5-4 contains the LocateErrors()
function that finds the roots of the error location polynomial specified in sigma[], takes the inverse of each root, and returns the error locations via error_vector[].

Listing 5-4

```
/*******************************/
/*                             */
/*   LocateErrors              */
/*                             */
/*******************************/
void LocateErrors(  exponent sigma[],
                    int bits_per_code_word,
                    bits error_vector[])
```

```
    {
exponent expon_value,k;

for(k=1;k<=bits_per_code_word;k++)
    {
    expon_value = EvalExtenFieldPoly(sigma,k);
    printf("for alpha**%d, expon_value = %d\n",k,expon_value);
    if(expon_value == -1)
        {
        error_vector[bits_per_code_word - k] = 1;
        }
    else
        {
        error_vector[bits_per_code_word - k] = 0;
        }
    }
return;
}
```

Convolutional Codes

6

Convolutional codes are a different breed than the block codes that have been covered thus far. Some authors contend that the two types of codes are **very** different, while other authors tend to emphasize their similarities. McEliece[1] shows how convolutional codes can be viewed as a special type of block code. The mathematical basis of convolutional coding is not as developed or rigorous as it is for block codes, and there are some fundamental obstacles that might cause this to always remain true. Nevertheless, convolutional codes are important weapons in the error-correction wars and they are used in many real-world systems.

6.1 Canonical Example

The easiest way to introduce convolutional codes is to start with a specific example.[2] Consider the structure shown in Fig. 6-1. Bits are shifted into the left end of the 3-stage shift register. The switch changes position at twice the input-shifting rate, alternately selecting the output of the top XOR gate and bottom XOR gate. The structure shown is an encoder for a single-input, 2-output, convolutional code with memory of length 3.

It is often useful to analyze a convolutional encoder as a state machine, and to represent the behavior of this machine in the form of a state diagram. The encoder of Fig. 6-1 has a memory of 3 bits, and upon casual observation, it seems as though it would therefore have $2^3 = 8$ possible states. This encoder can, in fact, be represented as a state machine with 8 states as we will demonstrate in Section 6.1.1. However, as you will subsequently discover in Section 6.1.2, the operation of this encoder can be represented using only four states.

---

1 R. J. McEliece: *Encyclopedia of Mathematics and Its Applications, Vol 3: The Theory of Information and Coding*, Addison-Wesley, Reading MA, 1977.

2 J. H. van Lint has dubbed this particular example the "canonical example" because it is used in almost every introduction to convolutional coding.

  J.H. van Lint: *Introduction to Coding Theory*, Springer-Verlag, Berlin, 1992.

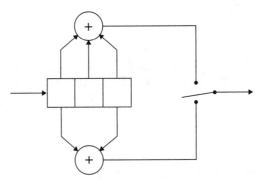

■ **6-1** *The so-called "canonical example" of a simple convolutional encoder.*

6.1.1 Convolutional Encoder Viewed as a Moore Machine

Start with the assumption that the encoder does have 8 possible states, with each state being defined by one of the eight possible combinations of 3 bits in the register. It's easy enough to associate a pair of outputs T_U, T_L with each state by realizing that

$$T_U = b_L + b_M + b_R$$
$$T_L = b_L + b_R$$

where

T_U = upper output
T_L = lower output
b_L = left bit in register
b_M = middle bit in register
b_R = right bit in register

Depending on its value, a new input bit will drive the machine from its old state to one of two possible new states. The new output is issued **after** arrival at the new state. The corresponding state diagram is shown in Fig. 6-2. Each state is represented by a circle containing its defining 3-bit combination and its corresponding two bits of output. A machine such as this one, in which the outputs are associated with the states, is called a *Moore machine*.

6.1.2 Convolutional Encoder Viewed as a Mealy Machine

Let's examine the first few steps in the operation of the example encoder to see how it might be possible to eliminate some steps:

1. Assume that at time 0, all three register stages contain zeros. The output bits are 00.

106

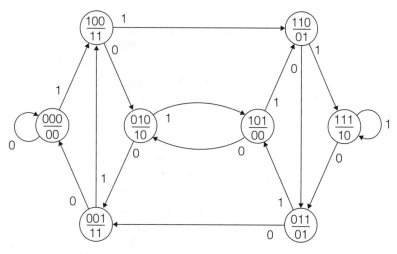

■ 6-2 *State diagram for the encoder of Fig. 6-1 represented as a Moore machine.*

2. The first input bit, u_1, is then shifted into the leftmost stage of the register. The register contents are $(u_1, 0, 0)$. The first output bit v_1 is obtained from the output of the upper summer. Therefore,

$$v_1 = u_1 + 0 + 0 = u_1$$

The second output bit, v_2, is obtained from the output of the lower summer as

$$v_2 = u_1 + 0 = u_1$$

The content of the leftmost stage is shifted into the middle stage, and second input bit, u_2, is shifted into the leftmost stage of the register. The register contents are now $(u_2, u_1, 0)$. The third output bit, v_3, is obtained from the output of the upper summer as

$$v_3 = u_2 + u_1 + 0$$

The fourth output bit, v_4, is obtained from the output of the lower summer as

$$v_4 = u_2 + 0$$

4. The content of the middle stage is shifted into the right stage, and the content of the left stage is shifted into the middle stage. The third input bit, u_3, is shifted into the left stage. The register contents are now (u_3, u_2, u_1). The fifth and sixth outputs are given by

$$v_5 = u_3 + u_2 + u_1$$
$$v_6 = u_3 + u_1$$

5. The content of the right stage is shifted out of existence, the content of the middle stage is shifted into the right stage, and the content of the left stage is shifted into the middle stage. The fourth input bit, u_4, is shifted into the left stage. The register contents are now (u_4, u_3, u_2). The seventh and eighth outputs are given by

$$v_7 = u_4 + u_3 + u_2$$
$$v_8 = u_4 + u_2$$

In each case, the outputs are determined by the three stages of the **new** register contents, or equivalently by the value of the input along with the **old** contents of the two leftmost stages of the register **just prior** to insertion of the new input bit. The insertion of the new input bit will push the contents of the rightmost stage completely out of the register. Therefore, two register states which differ only in the rightmost stage will yield the same next state for any given input. It seems as though we could redefine our sequential state machine in terms of just four 2-bit states, which are defined by the contents of the two leftmost stages of the register. The outputs, which are defined in terms of the original 3-bit states, must be mapped into the equivalent combinations of inputs and 2-bit states. This mapping is listed in Table 6-1.

■ Table 6-1 Mapping from 3-bit states to combinations of 2-bit states and inputs.

3-bit state	Output	2-bit state	Input
0 0 0	0 0	0 0	0
0 0 1	1 1	0 1	0
0 1 0	1 0	1 0	0
0 1 1	0 1	1 1	0
1 0 0	1 1	0 0	1
1 0 1	0 0	0 1	1
1 1 0	0 1	1 0	1
1 1 1	1 0	1 1	1

Each 2-bit state appears in the table twice, and each appearance has a different output. (For example, 2-bit state 01 appears once with output 11 for the 3-bit state 001 and once with output 00 for the 3-bit state 101.) Therefore, the outputs cannot be placed in the state circles as they are in Fig. 6-2. Instead, the outputs must be associated with the transition arrows, which are the only entities in the diagram that embody both input information and 2-bit state information. (Each transition arrow starts at the 2-bit state and is labeled with the input value that triggers the depicted transition.) A state diagram using 2-bit states is shown in Fig. 6-3. Each state is represented by a circle, labeled with the two leftmost register-bits that define the state. The transitions have two-part labels. The bit before the slash is the value of the input that will cause the indicated transition to occur. The two bits after the slash are the two output bits associated with the indicated transition. A machine, such as this one, in which the outputs are associated with the transition, is called a *Mealy* machine.

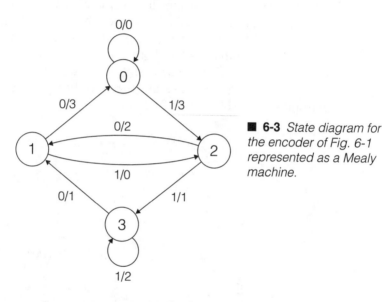

■ **6-3** *State diagram for the encoder of Fig. 6-1 represented as a Mealy machine.*

6.1.3 Moore Machines vs. Mealy Machines

It is always possible to convert a Mealy machine into an equivalent Moore machine and vice versa. The number of states in a Moore machine will equal or (usually) exceed the number of states in the equivalent Mealy machine. The Mealy machine of Fig. 6-3 can be converted into the Moore machine of Fig. 6-2 using a few rules of sequential machine design. First, a few definitions must be introduced to support the algorithm to follow.

Definition 6.1 A state of a Mealy machine is *z-homogeneous* if the output associated with each transition line incident (*i.e.*, pointing into) to that state is the same; otherwise, it is *z-nonhomogeneous*.

A z-nonhomogeneous state of a Mealy machine can be split into an equivalent set of z-homogeneous states.

Definition 6.2 A Mealy machine is *z-homogeneous* if every state of the machine is z-homogeneous; otherwise it is *z-nonhomogeneous*.

A z-homogeneous Mealy machine is equivalent to a Moore machine. Because the outputs associated with all the transitions incident to a z-homogeneous state are equal, the output can be associated directly with the state.

Example 6.6.1 Incident to state 10 in Fig. 6-3 are two transitions— one with output 00 and one with output 11. Hence, state 10 is z-nonhomogeneous. We can split state 10 into two states (for example, state 10z00 and state 10z11) as shown in Fig. 6-4. State 10z00 becomes the target for all transitions incident to state 10 with output 00, and state 10z11 becomes the target for all transitions incident to state 10 with output 11. All transitions out of state 10 must be duplicated for both states 10z00 and 10z11. (The transition labeled 0/10 going from state 10 to state 01 has been replaced with a transition from state 10z00 to state 01 plus a transition from state 10z11 to state 01.)

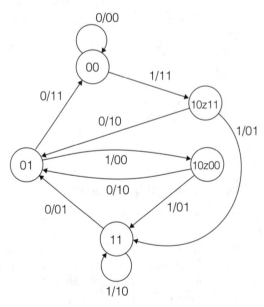

■ **6-4** *State diagram of a Mealy machine showing how z-nonhomogeneous state 10 can be split into two z-homogeneous states 10z11 and 10z00.*

Splitting states gets just a bit trickier for states that have *slings* (*i.e.*, transitions that start and end at the same state). In these cases, one of the new "post-split" states will retain the sling and there will be a transition from each of the other "post-split" states to the state that retains the sling. The input/output label on each of these transitions will match the input/output label on the sling.

Example 6.6.2 Incident to state 11 in Fig. 6-4 are three transitions—two with output 01 and one (a sling) with output 10. We can split state 11 into two states 11z01 and 11z10 as shown in Fig. 6-5.

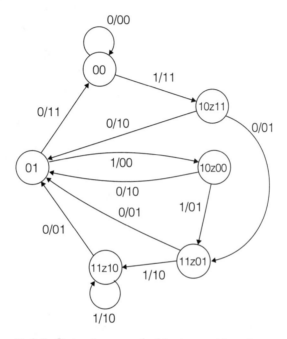

■ **6-5** *State diagram of a Mealy machine showing how a z-nonhomogeneous state with a sling (state 11) can be split into two z-homogeneous states 11z01 and 11z10.*

This is Example 1: Example 6.6.1. The results of Example 6.2 can be further extended by splitting state 00 to obtain Fig. 6-6 and then splitting state 01 to obtain Fig. 6-7, which is a z-homogeneous Mealy machine. Comparison of Figs. 6-7 and 6-2 reveals that the two machines are equivalent with the following correspondences between their states:

$$00z00 = 000$$
$$00z11 = 001$$
$$10z11 = 100$$

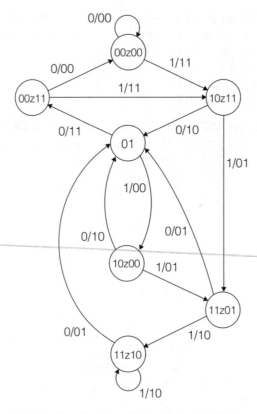

■ **6-6** *Result of splitting state 00 in the Mealy machine of Fig. 6-5.*

$$10z00 = 101$$
$$01z10 = 010$$
$$01z01 = 011$$
$$11z01 = 110$$
$$11z11 = 111$$

It often seems natural to think of the transitions as transient phenomena, which are short-lived relative to the possibly long dwell time that the machine spends within a state while waiting for an input to trigger the next transition. For this reason, texts on sequential machines often treat the outputs of a Mealy machine as *pulses* and the outputs of a Moore machine as *levels*. For purposes of encoder design, the outputs can be viewed as either pulses or levels.

6.1.4 Notation and Terminology

Notation and terminology for describing convolutional codes is not standard throughout the literature. This can be a source of con-

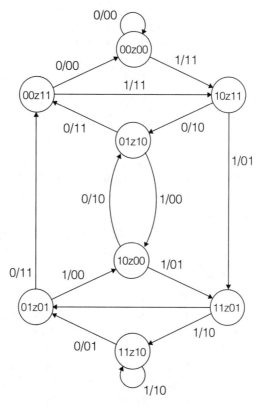

■ **6-7** *The z-homogeneous Mealy machine obtained by splitting state 01 in the Mealy machine of Fig. 6-6.*

siderable confusion for novices attempting to use multiple sources. Some of the confusion seems related to the choice between the Moore machine and Mealy machine for representing the encoder behavior.

Lin and Costello

The convention followed by Lin & Costello[3] and Rhee[4] allows the input to be "used" without first being shifted into a register. Thus, they would draw the encoder of Fig. 6-1 in the form shown in Fig. 6-8. This approach uses only 2 stages in the shift register and is perhaps more intuitively consistent with the 4-state Mealy machine of Fig. 6-3. Further bolstering this consistency is the fact that the outputs are shown directly as functions of **both** the 2-bit

3 S. Lin and D. J. Costello: *Error Control Coding: Fundamentals and Applications*, Prentice-Hall, Englewood Cliffs, NJ, 1983.

4 M. Y. Rhee: *Error Correcting Coding Theory*, McGraw-Hill, New York, 1989.

register state and the input. Both systems characterize convolutional codes using the ordered triple (n, k, m) where

$$n = \text{number of outputs}$$
$$k = \text{number of inputs}$$
$$m = \text{memory order or memory length}$$

In the code from Lin & Costello and Rhee, the *constraint length* n_A is defined as

$$n_A = n(m + 1)$$

The code implemented by Figs. 6-1 and 6-8 would be classified as $(2, 1, 2)$ code with a constraint length of 6. The connections to the shift register are specified by associating a polynomial with each of the outputs:

$$\text{upper XOR gate: } g^{(1)} = (1\ 1\ 1)$$
$$\text{lower XOR gate: } g^{(2)} = (1\ 0\ 1)$$

Clark and Cain

Clark and Cain[5] characterize a convolutional code in terms of the code rate, and implicitly assume that when the code rate is expressed as a common fraction, m/n, then m is the number of inputs and n is the number of outputs. They denote the number of stages as k and call this quantity the *constraint length*. [For codes with multiple registers, k is the total number of register stages.] This definition of constraint length is different than that of Lin & Costello and Rhee. Using the conventions of Clark and Cain, the code implemented by Fig. 6-1 would be characterized as a rate ½, constraint length 3 code.

Michelson and Levesque

Michelson and Levesque[6] structure their notation and presentation around the idea that each input shift can accept a multi-bit symbol rather than just a single bit. Each symbol shift is called a *cycle* and the following notation is used

$$b = \text{bits per cycle}$$
$$k = \text{number of register stages}$$
$$V = \text{number of outputs}$$

The number of register stages k is in units of symbol stages, *i.e.,* for $b > 1$, k is given by

$$k = \frac{total\ number\ of\ register\ bits}{b}$$

5 G. C. Clark and J. B. Cain: *Error-Correction Coding for Digital Communications*, Plenum Press, New York, 1981.

6 A. M. Michelson and A. H. Levesque: *Error-Control Techniques for Digital Communication*, Wiley-Interscience, New York, 1985.

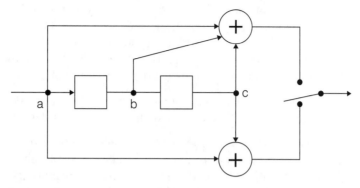

■ **6-8** *The canonical example encoder redrawn following the convention used by Lin and Costello.*

Constraint length is defined as the number of symbol registers k. (It should be noted that Michelson and Levesque warn their readers about alternative definitions for constraint length.)

Viterbi

Viterbi[7] defines the *constraint length* as the number of **bit** stages in the encoder shift register. The number of **symbol** stages is denoted by K, and the constraint length is bK, where b is the number of bits per symbol. For the binary case, $b = 1$, and the constraint length is simply K. Viterbi describes an encoder similar to Fig. 6-8 as having a 3-stage shift register, and hence the corresponding code is described as having a constraint length of 3. (The two boxes in Fig. 6-8 should be thought of as delay elements between the stages rather than as the stages of the shift register. The three stages are the three signal nodes labeled A, B, and C in the figure.)

Rather than associating a polynomial with each output, Viterbi associates a polynomial with each stage of the shift register.

$$\text{node a: } g_0 = (1, 1)$$
$$\text{node b: } g_1 = (1, 0)$$
$$\text{node c: } g_2 = (1, 1)$$

Summary

With the possible exception of k, it's not really a big deal which particular letters are used to denote specific parameters. However, it is very common for communications engineers to speak of a "rate ½, $k = 7$" convolutional code. These engineers (many of whom have never seen the Clark and Cain code) don't say "mem-

---

7 A. J. Viterbi and J. K. Omura: *Principles of Digital Communication and Coding*, McGraw-Hill, New York, 1979.

ory of 7" or "$m = 7$"; they say "$k = 7$." Apparently, because of just common usage, the notation used by Clark and Cain seems to have acquired a certain "preferred" status despite the fact that it disagrees with that of Lin and Costello, which is somewhat of a *de facto* "standard" with regard to other terminology.

In addition to the notation and terminology discussed so far, there are a number of items on which there is general agreement.

Just as for a block code, we can define the *code rate* as the ratio of information bits to total channel bits. Conceptually, the output of a convolutional encoder is semi-infinite, beginning at time zero and continuing forever. Therefore, the code rate R would simply be the ratio of encoder inputs to encoder outputs.

$$R = \frac{k}{n}$$

However, in the real world, the sequence of input bits will always be finite in length, and it will be necessary to append several zero bits to the input in order to "flush" the encoder registers and obtain all the output bits associated with the final information bit in the input sequence. This will increase slightly the percentage of noninformation bits in the encoder's output sequence. If the input sequence contains L bits and the encoder has m bits of memory, the *truncated code rate* R_T will be given by

$$R_T = \frac{kL}{n(m + L)} = R\left(\frac{L}{m + L}\right) = R\left(1 - \frac{m}{m + L}\right)$$

The quantity $m/(m + L)$ is sometimes referred to as the *fractional rate loss*.

6.2 Tree Representation of a Convolutional Encoder

The state diagram of an encoder is a compact way to depict all of the possible states and transitions between them, but it does not provide a good view of the encoder operation as being sequential in time. Let's try to adapt the state diagram into something which is better able to depict the sequential nature of encoder operation. Consider the state diagram in Fig. 6-3. Assume that at $t = 0$ the encoder is in state 00. For $t = 1$, only two states are possible: the encoder will remain in state 00 if the input is 0, or the encoder will change to state 10 if the input is 1. We can illustrate this fact by drawing just a portion of the state diagram as shown in Fig. 6-9. Notice how the states for $t = 1$ have been placed in a vertical column directly above the notation "$t = 1$." Now for transitions from $t = 1$ to $t = 2$, we must consider two cases. If the encoder is in state

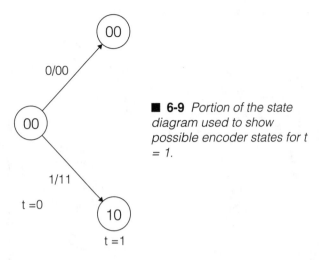

■ **6-9** *Portion of the state diagram used to show possible encoder states for t = 1.*

00 at $t = 1$, it can be in either state 00 or state 10 at $t = 2$. (Which one depends upon the input value.) Similarly, if the encoder is in state 10 at $t = 1$, it can change to either state 01 or state 11 at $t = 2$. We can illustrate these possibilities by adding another column to the diagram of Fig. 6-9 to obtain the diagram shown in Fig. 6-10. We could repeat the process for transitions from $t = 2$ to $t = 3$, and obtain the diagram shown in Fig. 6-11. It's easy to see how this diagram could quickly become unwieldy for even moderate values of

117

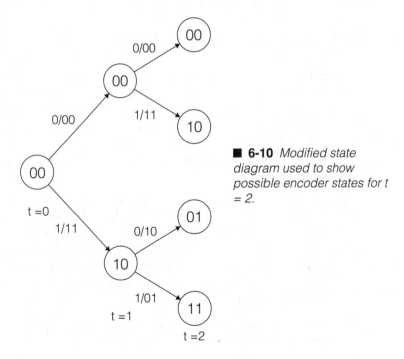

■ **6-10** *Modified state diagram used to show possible encoder states for t = 2.*

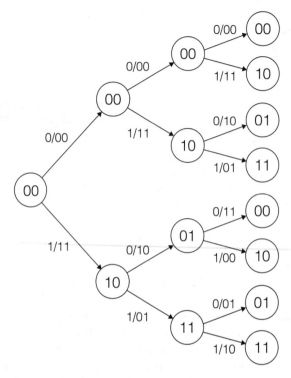

■ **6-11** *Modified state diagram used to show possible encoder states for t = 3.*

t. Several simplifications are usually made to make the diagram more compact:

1. In Fig. 6-11 each state is represented by a circle with one transition incident from the left and two transitions leaving from the right. This can be simplified by eliminating the circle and allowing the state to be implicitly represented by the intersection of the three transition lines. (If the input is not binary, there will be more than two transitions leaving from each state.)

2. The transitions leaving to the right of each state can be arranged in order of input value so that it becomes unnecessary to label each transition with the input value that triggers it. In most of the literature, the transition corresponding to a 0 input is drawn above the transition corresponding to a 1 input for binary encoders, but in some cases[8] the order is reversed.

8 A. M. Michelson and A. H. Levesque: *Error-Control Techniques for Digital Communication*, Wiley-Interscience, New York, 1985.

The result of these simplifications is called a *tree*. A tree for the encoder of Fig. 6-1 is shown in Fig. 6-12. At each fork, the number of branches equals the number of possible values for a single input symbol. The bits superimposed on each branch are the outputs issued in response to the input values corresponding to the branch.

Example 6.6.3 Assume that the encoder register contains all zeros prior to the arrival of the first input bit. The sequence of input bits is 1, 0, 0, 1, 0. Figure 6-13 shows the tree of Fig. 6-12 with the path corresponding to this input sequence highlighted. Following along the highlighted path we find the output sequence 11, 10, 11, 11, 10.

Figure 6-12 shows a tree the way it usually appears in the literature. I have found that it is sometimes easier to keep things straight if each fork is annotated with the encoder state as in Fig. 6-14.

6.3 Trellis Representation of a Convolutional Encoder

Even with simplifications, a tree can become difficult to draw as t increases. For binary encoders, each state will appear $2^{\tau-2}$ times in the column for $t = \tau$. This amounts to unnecessary repetition because for each appearance of a particular state, the output and next state will always be the same for a given input.

Example 6.6.4 Examination of Fig. 6-12 reveals that:

1. State 00 always reacts to an input of 0 by issuing 00 as output and remaining in state 00.

2. State 00 always reacts to an input of 1 by issuing 11 as output and changing to state 10.

3. State 01 always reacts to an input of 0 by issuing 11 as output and changing to state 00.

4. State 01 always reacts to an input of 1 by issuing 00 as output and changing to state 10.

5. State 10 always reacts to an input of 0 by issuing 10 as output and changing to state 01.

6. State 10 always reacts to an input of 1 by issuing 01 as output and changing to state 11.

7. State 11 always reacts to an input of 0 by issuing 01 as output and changing to state 01.

8. State 11 always reacts to an input of 1 by issuing 10 as output and remaining in state 10.

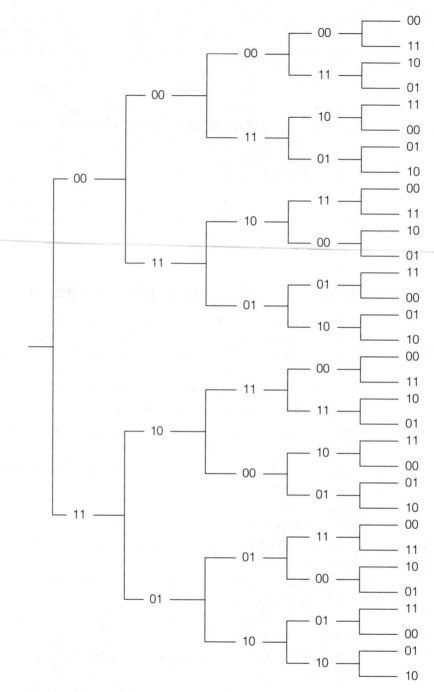

■ **6-12** *A tree used to depict possible sequences of encoder states for the encoder of Fig. 6-1.*

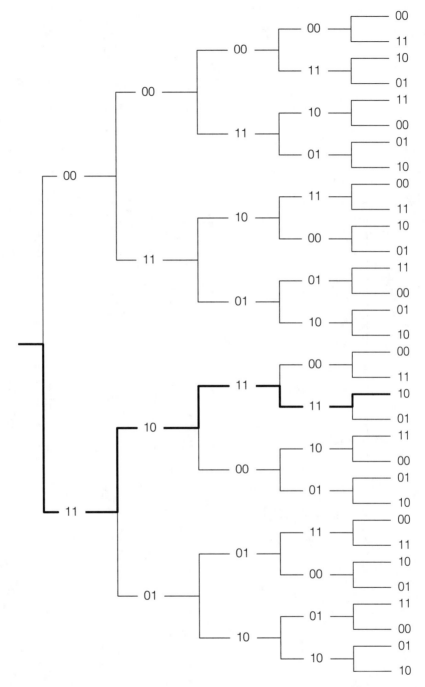

■ **6-13** *An encoder tree with path highlighted for the input sequence 10010.*

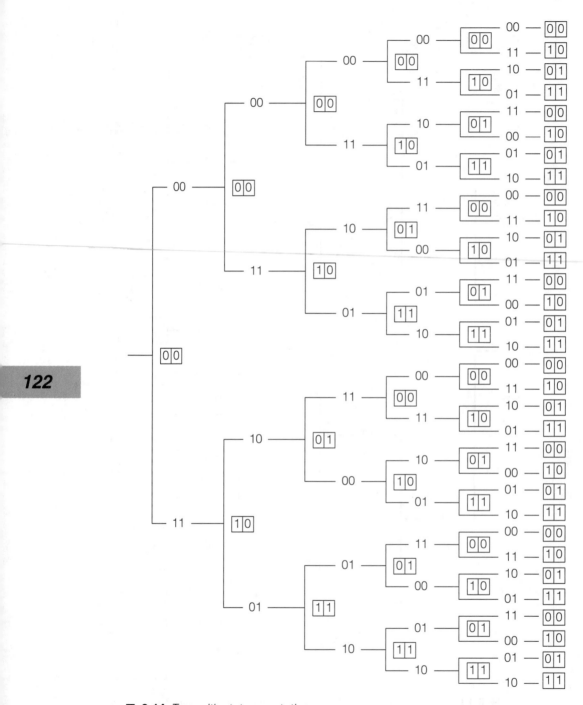

■ **6-14** *Tree with state annotations.*

A tree can be turned into a *trellis* by allowing each state to appear only once in each column. Because of the fact that a given input will always produce the same output and next state for each appearance of a given state, it will always be possible to draw the necessary transitions between columns of the trellis without creating conflicts or ambiguities. The trellis corresponding to Fig. 6-12 is shown in Fig. 6-15. There are several different approaches for keeping track of which transitions are caused by which inputs. Figure 6-15 follows the same convention used in drawing trees—the states are arranged within each column so that for the transitions **leaving** a state, the transition caused by a 0 input is drawn above the transition because of a 1 input. For some codes, it might be difficult or impossible to find an order for the states that will allow this convention to be used. For binary codes, many authors use solid lines to draw transitions corresponding to 0 inputs and dashed lines to draw transitions corresponding to 1 inputs as shown in Fig. 6-16. It might be possible to extend this convention to nonbinary codes with small symbol alphabets by using different types of broken lines such as small dashes, long dashes, dashes alternated with dots, etc. Although it leads to more cluttered diagrams, the one approach guaranteed to work for all codes is to simply retain the two-part "input/output" label used in state diagrams, as shown in Fig. 6-17.

6.4 Distance Measures

In the analysis of block codes, the minimum distance of a code proves to be an important measure for assessing the performance

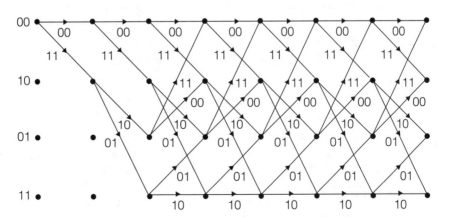

■ **6-15** *A trellis used to depict possible sequences of encoder states for the encoder of Fig. 6-1.*

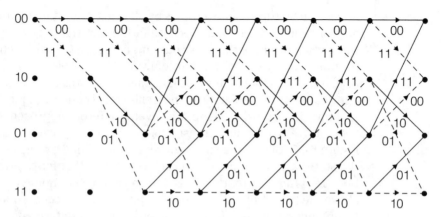

■ 6-16 *Trellis with dashed lines used to draw transitions corresponding to input = 1.*

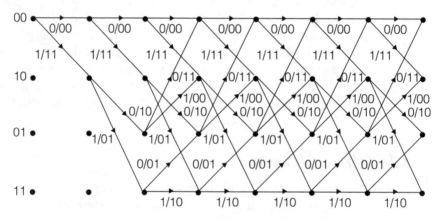

■ 6-17 *Trellis having each transition labeled with both input and output.*

of the code. The same is true of convolutional codes, but the definition of "minimum distance" needs to be adjusted slightly for the different nature of convolutional codes.

For convolutional codes, the minimum distance between any two codewords is called the *minimum free distance* and is denoted as d_{free}. Because a convolutional code is a linear code, we can take one of the codewords as the all-zero codeword and view the minimum free distance as the weight of the minimum-weight nonzero codeword. Consider the code defined by the state diagram of Fig. 6-3. Because all codewords begin and end with the encoder at state 00, each nonzero codeword corresponds to sequence of state transitions that departs from state 00 for some number of state times before returning to state 00. Thus, the minimum-weight nonzero codeword corresponds to the minimum-weight path that departs from and returns to state 00. The encoder defined by Fig. 6-3 is simple enough that we can find such a minimum-weight path by inspection. The only way out of state 00 is via the transition to state 10. This transition has a weight of 2. Similarly, the only way back into state 00 is via the transition from state 01. This transition also has a weight of 2. The transition from state 10 to state 01 has a weight of 1. The minimum-weight path that departs from and returns to state 00 goes through states 10 and 01 and has a total weight of $2 + 1 + 2 = 5$.

Viterbi Decoding

7

In 1967, Viterbi[1] introduced a decoding procedure for convolutional codes that was subsequently shown to provide a maximum likelihood decoding of convolutional codes. The Viterbi algorithm or Viterbi decoder (as it has come to be called), is currently the most widely used technique for the decoding of convolutional codes. This section develops the basic algorithm by means of an example decoding operation.

7.1 Introduction to Viterbi Decoding

Example 7.1 Once again, let's consider the canonical example presented in Section 6.1. The state diagram for the encoder is reproduced here as Fig. 7-1. Assume that each codeword is produced from 7 information bits plus a 2-bit tail of zeros added to flush out the encoder. The encoder trellis as tailored to this specific case is shown in Fig. 7-2. (Because the eighth and ninth input bits will always be zeros, the transitions corresponding to inputs of 1 are eliminated from the eighth and ninth transition columns of the trellis.) Now suppose that the information to be transmitted is 1010001, the 7-bit ASCII code for Q. We traverse the encoder trellis via the path shown in Fig. 7-3 to determine that the resulting 18-bit codeword is

$$11\ 10\ 00\ 10\ 11\ 00\ 11\ 10\ 11$$

The signal is transmitted over a binary symmetric channel (BSC). Now suppose that two bit errors occur so that the received codeword is

$$10\ 10\ 00\ 10\ 11\ 01\ 11\ 10\ 11$$

How should we proceed with decoding in order to correct the received errors and determine the transmitted information sequence? Let's start by considering just the first two bits of the

---

1 A. J. Viterbi, "Error Bounds for Convolutional Codes and an Asymptotically Optimum Decoding Algorithm," *IEEE Trans. Inf. Theory*, IT-13, pp. 260–269, April 1967.

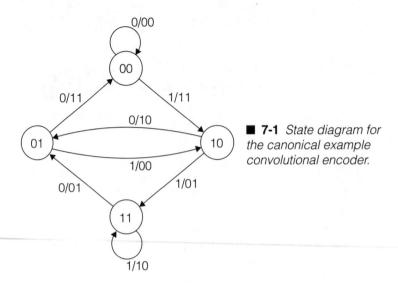

■ **7-1** *State diagram for the canonical example convolutional encoder.*

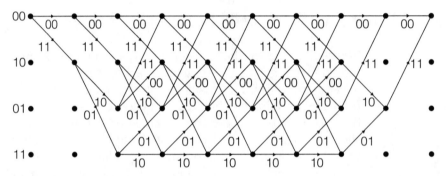

■ **7-2** *Trellis of the canonical example as modified for the specific case of 7 information bits plus 2-bit tail.*

input: 1 0 1 0 0 0 1 0 0
output: 11 10 00 10 11 00 11 10 11

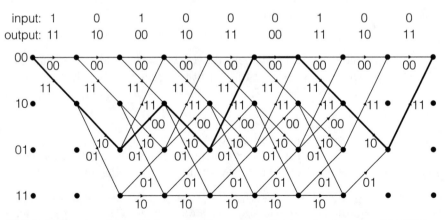

■ **7-3** *Trellis showing path traversed to encode the input sequence 1010001.*

128

received sequence. Examination of the first transition column of the trellis reveals that a received 10 could have occurred in one of two ways:

1. The first uncoded information bit was 1, which produced an encoder output symbol of 11. This symbol was received (with an error in the second bit) as 10.

2. The first uncoded information bit was 0, which produced an encoder output symbol of 00. This symbol was received (with an error in the first bit) as 10.

In a BSC, either of these two possibilities is equally likely, so we really have no basis for choosing one over the other. Therefore, we will keep both possibilities under consideration. We can depict our decoding efforts so far in the form of the partial tree shown in Fig. 7-4. Each branch is labeled with the encoder output symbol plus a second number called the *branch metric*. The precise rule for determining a branch metric varies depends upon the particular application, but in all cases it is a measure of relative "goodness" among the various branches. In the present example, the branch metric is equal to the Hamming distance between the received bits and the encoder output for that particular branch. Another way to look at this is that the branch metric is the number of bit errors that would have had to occur in order to produce the received bits given the particular encoder transition. Each of the two branches has an output symbol that differs from the received symbol by one bit, so each bit is labeled with 1. For this metric based on Hamming distance, smaller metric values are associated with the "better" branches. As is covered further on, there are other metrics in which the larger values are associated with the "better" branches.

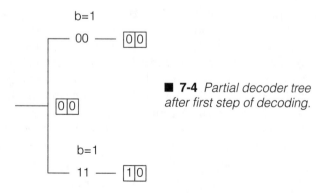

■ **7-4** *Partial decoder tree after first step of decoding.*

Let's continue by considering the next two received bits, which are 10. The received sequence 1010 could have occurred in one of four possible ways:

1. The first uncoded information bit was 1, and the second uncoded information bit was 1. Together these inputs produced an encoder output sequence 1101. Three bit errors occurred (in bits 2, 3, and 4), causing the transmitted sequence 1101 to be received as 1010.

2. The first uncoded information bit was 1, and the second uncoded information bit was 0. Together these inputs produced an encoder output sequence 1110. One bit error occurred (in bit 2), causing the transmitted sequence 1110 to be received as 1010.

3. The first uncoded information bit was 0, and the second uncoded information bit was 1. Together these inputs produced an encoder output sequence 0011. Two bit errors occurred (in bits 1 and 4), causing the transmitted sequence 0011 to be received as 1010.

4. The first uncoded information bit was 0, and the second uncoded information bit was 0. Together these inputs produced an encoder output sequence 0000. Two bit errors occurred (in bits 1 and 3), causing the transmitted sequence 0000 to be received as 1010.

These possibilities are depicted in Fig. 7-5. Now, each transition in the second column is labeled with two values: the *branch metric* b and the *path metric* p. The path metric is the sum of the branch metrics for all the transitions in the path from the original state to the indicated point.

At this point, a strategy for brute-force decoding should be apparent. Simply compute the path metric for every path through the tree, and select the path with the smallest metric as the decoder's estimate of the correct path. The decoded information bits would be deemed equal to those bits needed to cause the selected path to be traversed. This **seems** like a viable approach, but it quickly grows unwieldy as additional symbols are processed. Let's see how the tree grows as we process an additional received symbol. Examination of the encoder trellis reveals that received sequence 10 10 00 could have occurred in one of eight ways:

1. The first uncoded information bit was 0, the second uncoded information bit was 0, and the third uncoded information bit was 0. Together these inputs produced an encoder output sequence 000000. Two bit errors occurred (in bits 1 and 3),

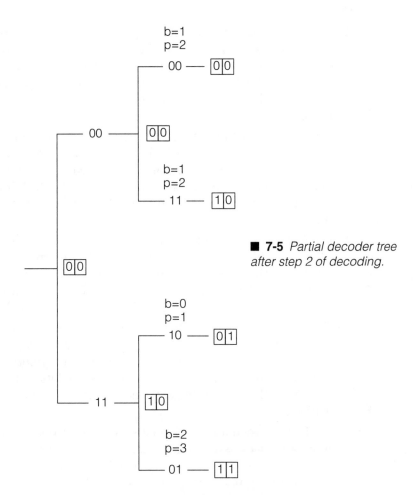

131

■ **7-5** *Partial decoder tree after step 2 of decoding.*

causing the transmitted sequence 000000 to be received as 101000.

2. The first uncoded information bit was 0, the second uncoded information bit was 0, and the third uncoded information bit was 1. Together these inputs produced an encoder output sequence 000011. Four bit errors occurred (in bits 1, 3, 5, and 6), causing the transmitted sequence 000011 to be received as 101000.

3. The first uncoded information bit was 0, the second uncoded information bit was 1, and the third uncoded information bit was 0. Together these inputs produced an encoder output sequence 001110. Three bit errors occurred (in bits 1, 4, and 5), causing the transmitted sequence 001110 to be received as 101000.

4. The first uncoded information bit was 0, the second uncoded information bit was 1, and the third uncoded information bit was 1. Together these inputs produced an encoder output sequence 001101. Three bit errors occurred (in bits 1, 4, and 6), causing the transmitted sequence 001101 to be received as 101000.

5. The first uncoded information bit was 1, the second uncoded information bit was 0, and the third uncoded information bit was 0. Together these inputs produced an encoder output sequence 111011. Three bit errors occurred (in bits 2, 5, and 6), causing the transmitted sequence 111011 to be received as 101000.

6. The first uncoded information bit was 1, the second uncoded information bit was 0, and the third uncoded information bit was 1. Together these inputs produced an encoder output sequence 111000. One bit error occurred (in bit 2), causing the transmitted sequence 111000 to be received as 101000.

7. The first uncoded information bit was 1, the second uncoded information bit was 1, and the third uncoded information bit was 0. Together these inputs produced an encoder output sequence 110101. Four bit errors occurred (in bits 2, 3, 4, and 6), causing the transmitted sequence 110101 to be received as 101000.

8. The first uncoded information bit was 1, the second uncoded information bit was 1, and the third uncoded information bit was 1. Together these inputs produced an encoder output sequence 110110. Four bit errors occurred (in bits 2, 3, 4, and 5), causing the transmitted sequence 110110 to be received as 101000.

As shown in Fig. 7-6, there are eight possible paths. If we were to continue with the fourth pair of received bits, we would find 16 possible paths. In general, at level N in the tree, there would be 2^N paths. The value of N doesn't have to get very large before the number of paths (and hence the number of path-metric computations) becomes prohibitive. For $N = 100$, the number of paths becomes $2^N \cong 1.27 \times 10^{30}$. Assuming each path-metric computation takes 1 microsecond, the total calculation would take 40 quadrillion years!

Let's take another look at the partial decode of Fig. 7-6 by collapsing the tree into a trellis as shown in Fig. 7-7. (The number located near the terminal end of each branch is the cumulative path metric for the path that ends with the particular branch.) State 00 at

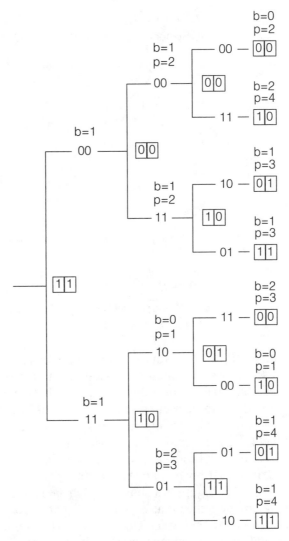

■ **7-6** *Partial decoder tree after step 3 of decoding.*

level 3 in the trellis can be reached by two different paths—one with a path metric of 2 (state sequence 00,00,00,00) and one with a path metric of 3 (state sequence 00,10,01,00). No matter what happens in subsequent levels, any complete path that passes through state 00 at level 3 will include one or the other of these two partial paths from state 00 level 0 to state 00 level 3. Therefore, if we were to compare two complete paths that differ only in their first three branches, the path metric for the path that starts off with states 00, 10, 01, 00 will always be larger by 1 than the

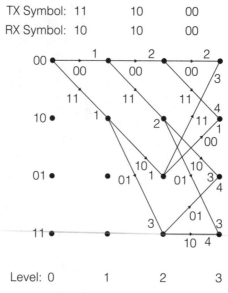

TX Symbol: 11   10   00

RX Symbol: 10   10   00

Level: 0   1   2   3

■ **7-7** *Partial trellis obtained by collapsing partial tree of Fig. 7-6.*

path metric of the path that starts off with states 00,00,00,00. Because we are ultimately interested in finding the complete path with the smallest overall metric, why bother even considering paths that are guaranteed never to be the smallest? At state 00 level 3, we could "kill off" the incident path with metric 3 and only build upon the surviving path, which has a path metric of 2. Likewise, at state 10 level 3, we could kill off the incident path with metric 4 and only build upon the surviving path, which has a path metric of 1. Figure 7-8 shows the partial trellis redrawn to eliminate any branches that do not participate in a surviving path. At each state of each level, we only keep the one incident path that could possibly participate in the ultimate "winning" path that has the smallest total path metric. Now at each state of each level, there will only be two incident paths for which to compute a path metric. Instead of 40 quadrillion years, the computations for level 100 would only take 8 microseconds.

Let's continue decoding of the fourth pair of received bits by building upon only the four **surviving** paths at level 3 in the trellis. The received sequence 10100010 could have occurred in one of eight ways:

1. The first three bits of the information sequence were 000, which leads to state 00 at level 3 in the trellis via a path

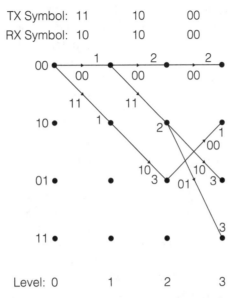

TX Symbol: 11   10   00

RX Symbol: 10   10   00

Level: 0   1   2   3

■ **7-8** *Partial trellis of Fig. 7-7 modified to show only the "surviving" paths.*

through states 00,00,00,00. The corresponding encoder output sequence would be 000000, which compared to the received sequence 101000 indicates two bit errors (in bits 1 and 3). Therefore, the partial path metric to this point is 2.

A. The fourth bit of the information sequence was 0, which causes a transition from state 00 at level 3 to state 00 at level 4. The corresponding encoder output bits would be 00, which compared to the fourth pair of received bits (10) indicates one bit error. Therefore, the transition from state 00 level 3 to state 00 level 4 has a branch metric of 1. The path metric for the path through states 00,00,00,00,00 is then 2 + 1 = 3.

B. The fourth bit of the information sequence was 1, which causes a transition from state 00 at level 3 to state 10 at level 4. The corresponding encoder output bits would be 11, which compared to the fourth pair of received bits (10) indicates one bit error. Therefore, the transition from state 00 level 3 to state 10 level 4 has a branch metric of 1. The path metric for the path through states 00, 00, 00, 00, 10 is then 2 + 1 = 3.

2. The first three bits of the information sequence were 101, which leads to state 10 at level 3 in the trellis via a path through states 00, 10, 01, 10. The corresponding encoder

output sequence would be 111000, which compared to the received sequence 101000 indicates one bit error (in bit 2). Therefore, the partial path metric to this point is 1.

A. The fourth bit of the information sequence was 0, which causes a transition from state 10 at level 3 to state 01 at level 4. The corresponding encoder output bits would be 10, which compared to the fourth pair of received bits (10) indicates no bit errors. Therefore, the transition from state 10 level 3 to state 01 level 4 has a branch metric of 0. The path metric for the path through states 00, 10, 01, 10, 01 is then $1 + 0 = 1$.

B. The fourth bit of the information sequence was 1, which causes a transition from state 10 at level 3 to state 11 at level 4. The corresponding encoder output bits would be 01, which compared to the fourth pair of received bits (10) indicates two bit errors. Therefore, the transition from state 10 level 3 to state 11 level 4 has a branch metric of 2. The path metric for the path through states 00, 10, 01, 10, 11 is then $1 + 2 = 3$.

3. The first three bits of the information sequence were 010, which leads to state 01 at level 3 in the trellis via a path through states 00,00,10,01. The corresponding encoder output sequence would be 001110, which compared to the received sequence 101000 indicates three bit errors (in bits 1, 4, and 5). Therefore, the partial path metric to this point is 3.

A. The fourth bit of the information sequence was 0, which causes a transition from state 01 at level 3 to state 00 at level 4. The corresponding encoder output bits would be 11, which compared to the fourth pair of received bits (10) indicates one bit error. Therefore, the transition from state 01 level 3 to state 00 level 4 has a branch metric of 1. The path metric for the path through states 00, 10, 01, 01, 00 is then $3 + 1 = 4$.

B. The fourth bit of the information sequence was 1, which causes a transition from state 01 at level 3 to state 10 at level 4. The corresponding encoder output bits would be 00, which compared to the fourth pair of received bits (10) indicates one bit error. Therefore, the transition from state 01 level 3 to state 10 level 4 has a branch metric of 1. The path metric for the path through states 00, 10, 01, 01, 00 is then $3 + 1 = 4$.

4. The first three bits of the information sequence were 011, which leads to state 11 at level 3 in the trellis via a path through states 00, 00, 10, 11. The corresponding encoder

output sequence would be 001101, which compared to the received sequence 101000 indicates three bit errors (in bits 1, 4, and 6). Therefore, the partial path metric to this point is 3.

A. The fourth bit of the information sequence was 0, which causes a transition from state 11 at level 3 to state 01 at level 4. The corresponding encoder output bits would be 01, which compared to the fourth pair of received bits (10) indicates two bit errors. Therefore, the transition from state 11 level 3 to state 01 level 4 has a branch metric of 2. The path metric for the path through states 00, 00, 10, 11, 01 is then 3 + 2 = 5.

B. The fourth bit of the information sequence was 1, which causes a transition from state 11 at level 3 to state 11 at level 4. The corresponding encoder output bits would be 10, which compared to the fourth pair of received bits (10) indicates no bit errors. Therefore, the transition from state 11 level 3 to state 11 level 4 has a branch metric of 0. The path metric for the path through states 00, 00, 10, 11, 01 is then 3 + 0 = 3.

This set of possibilities is depicted in Fig. 7-9. Notice that the two paths incident to state 11 at level 4 are "tied" with each of their path metrics equal to 3. In case of ties of this sort, we arbitrarily select either path as the survivor. The remaining steps of the decoding process are depicted in Figs. 7-10 through 7-14. After the fi-

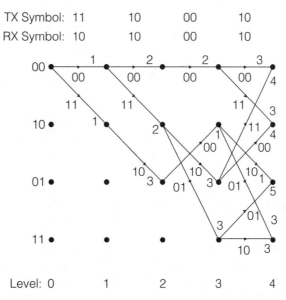

7-9 *Partial trellis after step 4 of decoding.*

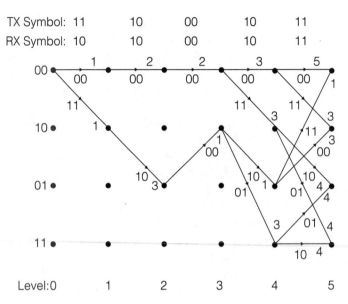

7-10 *Partial trellis after step 5 of decoding.*

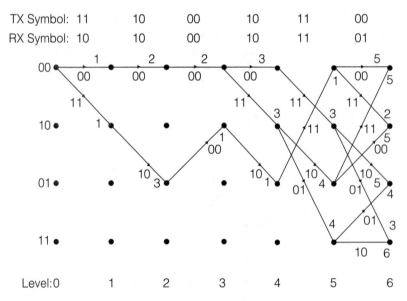

7-11 *Partial trellis after step 6 of decoding.*

nal step in the trellis building process, we simply select the path with the lowest total path metric. The sequence of input bits associated with the branches along this path constitutes the output of the decoder. The output corresponding to the highlighted path in Fig. 7-14 is 101000100, which exactly matches the encoder input sequence.

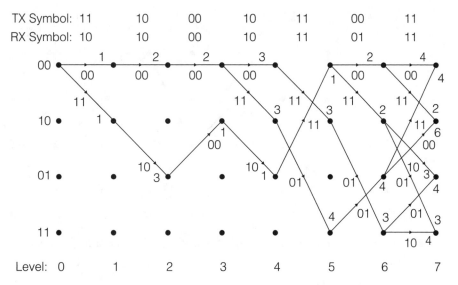

■ 7-12 *Partial trellis after step 7 of decoding.*

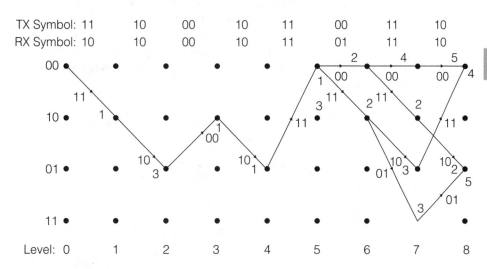

■ 7-13 *Partial trellis after step 8 of decoding.*

Redrawing the trellis after each step, as we have been doing in Example 7.1, helps eliminate the clutter that can make it hard to see exactly what is going on in the decoding process, but all the drawing can become tedious. An alternative to redrawing after each step is to simply put an X on each incident branch that is to be "killed-off," and use the cumulative path metric of the surviving incident branch to compute the path metrics for each branch leav-

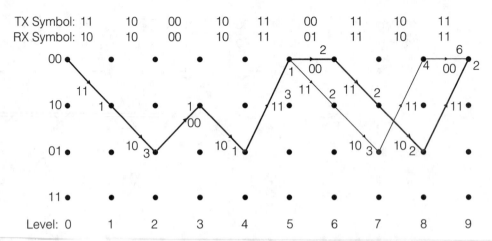

■ **7-14** *Partial trellis after final step of decoding. The path with the smallest metric is highlighted.*

ing a given state. This approach allows the drawing for each step to be obtained by simply adding on to the drawing from the previous step—it is not necessary to redraw, because once drawn, none of the branches are removed. Figures 7-15, 7-16, and 7-17 show this approach's results corresponding to Figs. 7-8, 7-9, and 7-14. Notice that Fig. 7-16 is a subset of 7-17, and Fig. 7-15 is a subset of both Figs. 7-16 and 7-17.

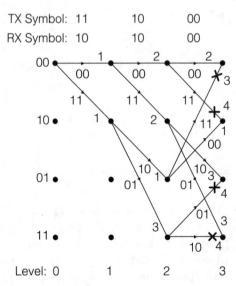

■ **7-15** *Alternative form for the partial trellis of Fig. 7-8.*

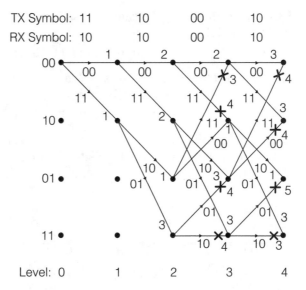

TX Symbol: 11   10   00   10
RX Symbol: 10   10   00   10

7-16 *Alternative form for the partial trellis of Fig. 7-9.*

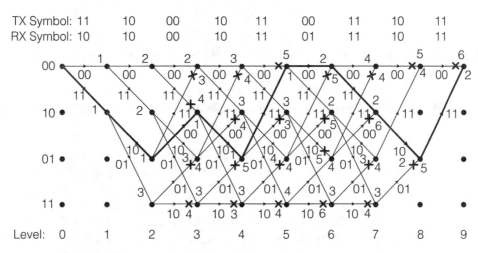

TX Symbol: 11 10 00 10 11 00 11 10 11
RX Symbol: 10 10 00 10 11 01 11 10 11

7-17 *Alternative form for the partial trellis of Fig. 7-14.*

141

7.2 Viterbi Decoding Failures

The Basic Idea

Viterbi decoding does not work equally well with all possible convolutional codes. A knowledge of the mechanisms involved in Viterbi decoding failures can be used to design or select codes that perform well under Viterbi decoding.

Example 7.2 Start with the same transmitted sequence considered in Example 7.1:

$$11 \ 10 \ 00 \ 10 \ 11 \ 00 \ 11 \ 10 \ 11$$

However, instead of just 2 errors, let's assume that five errors have occurred so that the received bit sequence is

$$10 \ 10 \ 11 \ 10 \ 11 \ 01 \ 11 \ 11 \ 11$$

The decoder trellis for this case is shown in Fig. 7-18. The decoder output corresponding to the highlighted path is 100011100, which differs from the original encoder input sequence in bits 3, 5, and 6.

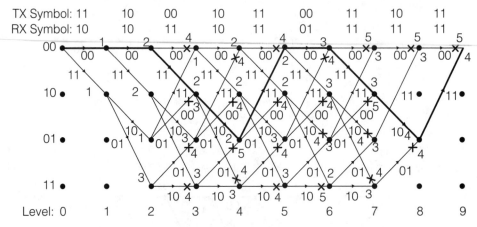

■ **7-18** *Trellis for case of incorrect decoding when there are 5 bit errors in the channel.*

The decoder trellis shown in Fig. 7-18 contains a tie at state 01 after step 8. Both incident branches at state 01 have a cumulative path metric of 4, and we arbitrarily select the upper branch as the survivor. We might be led to wonder about how the decoder might have performed if we had selected the lower incident branch instead. The resulting trellis for this case is shown in Fig. 7-19. The decoder output corresponding to the highlighted path is 100011100, which differs from the original encoder input sequence in bits 3, 5, and 6.

The encoder trellis of Fig. 7-3 can be inverted or "turned inside-out" to produce the equivalent trellis shown in Fig. 7-20. The states at each level in the trellis have been rearranged so that all of the branches in the encode path for the codeword 111000101100111011 lie along the top edge of this trellis. The state labels are enclosed in boxes. Because the states have been rearranged, it is no longer possible to rely on the upper/lower branch convention to indicate

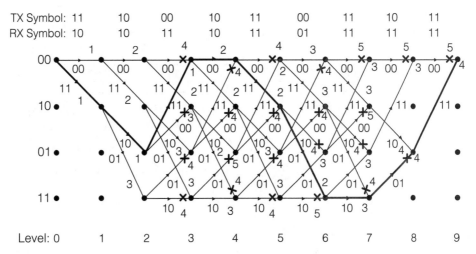

■ **7-19** *Modification of Fig. 7-18 based on alternate resolution of tie at state 01 after step 8.*

which branches correspond to inputs of 0 and 1. Therefore, the branches corresponding to a 0 input are drawn with a solid line, and branches corresponding to a 1 input are drawn with a dashed line.

If we draw the incorrect decode path of Fig. 7-18 on a trellis using the new state arrangement of Fig. 7-20, we obtain the result shown in Fig. 7-21. As stated previously, the received bit sequence is 10

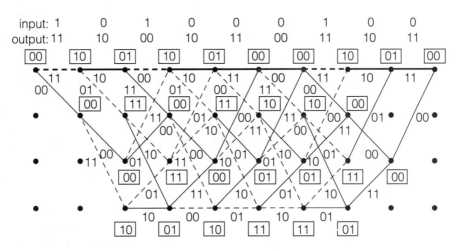

■ **7-20** *Trellis of Fig. 7-3 turned inside-out to place encode path along top edge.*

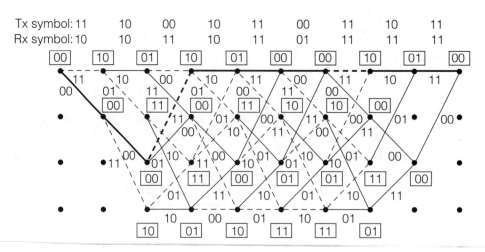

Tx symbol: 11   10   00   10   11   00   11   10   11
Rx symbol: 10   10   11   10   11   01   11   11   11

■ 7-21 *Case of incorrect decoding from Fig. 7-18 shown on inside-out trellis structure from Fig. 7-20.*

10 11 10 11 01 11 11 11. The error sequence is obtained as the bit by bit XOR of the transmitted sequence and the received sequence:

Tx	11	10	00	10	11	00	11	10	00
Rx	10	10	11	10	11	01	11	11	11
error	01	00	11	00	00	01	00	01	00

If we use the original trellis arrangement of Fig. 7-2 and decode this error sequence as though it were the received signal, we obtain the result shown in Fig. 7-22 (the tie at level 8 is resolved in favor of the upper branch just as it was for Fig. 7-18). The shape of the highlighted decode path is the same as the shape of the highlighted path in Fig. 7-21. Similarly, if we draw the incorrect decode path of Fig. 7-19 on a trellis using the new state arrangement of Fig. 7-20, we obtain the result shown in Fig. 7-23. If we then decode the error sequence again, this time resolving the tie at level 8 in favor of the lower branch, we obtain the result shown in Fig. 7-24. The shape of the highlighted decode path is the same as the shape of the highlighted path in Fig. 7-23. Conceptually, we could repeat this process for every possible combination of codeword and error sequence. As a consequence of the code's linearity, we would find that for each such combination decoded using an "inverted" trellis, the shape of the decode path would match the shape of the decode path obtained by decoding the error sequence alone using the original trellis. What is the significance of this fact? If we were to try and construct an "inverted" trellis for the all-zero codeword, the result would be identical to Fig. 7-2. When analyzing the performance of a convolutional code, it is not necessary to consider all

144

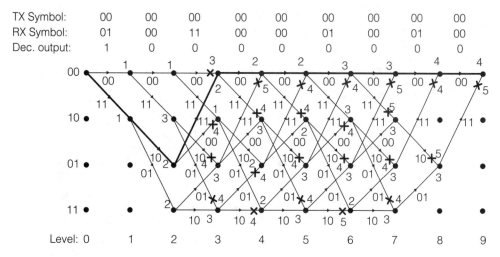

TX Symbol: 00   00   00   00   00   00   00   00   00
RX Symbol: 01   00   11   00   00   01   00   01   00
Dec. output: 1   0   0   0   0   0   0   0   0

■ **7-22** *Decoding the error sequence for Example 7.2 on the original trellis structure of Fig. 7-2. Tie at level 8 resolved in favor of the upper incident branch.*

possible codewords. For each decoding failure of a nonzero codeword there is an equivalent or *homomorphic* failure of the all-zero codeword. Thus, if we analyze all the ways in which the all-zero codeword might be incorrectly decoded, we will effectively be analyzing all the ways in which any codeword might be incorrectly decoded. This fact greatly simplifies the task of performance analysis.

Let's take another look at Figs. 7-22 and 7-24. In each case, the *error event* begins with a single branch that causes the decode path

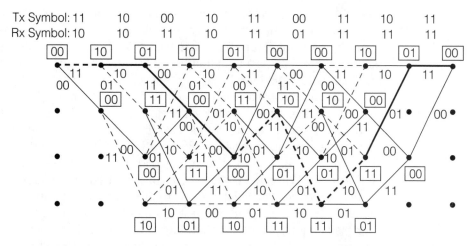

■ **7-23** *Case of incorrect decoding from Fig. 7-19 shown on inside-out trellis structure from Fig. 7-20.*

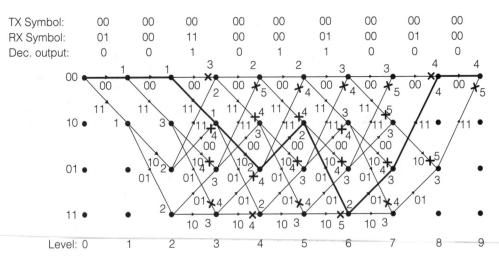

TX Symbol: 00   00   00   00   00   00   00   00   00
RX Symbol: 01   00   11   00   00   01   00   01   00
Dec. output: 0   0   1   0   1   1   0   0   0

■ **7-24** *Decoding the error sequence for Example 7.2 on the original trellis structure of Fig. 7-2. Tie at level 8 resolved in favor of the lower incident branch.*

to stray from the all-zero path. In Fig. 7-22, the path transitions to state 10 at level 1, and in Fig. 7-24, the path transitions to state 10 at level 3. In neither case can the path immediately return to state 00—there are no transitions going directly from state 10 to state 00.

7.2.1 Minimum Free Distance

Recall from Chapter 9, that the *minimum free distance* of a code is equal to the weight of the minimum-weight path that departs from state 00 for some number of state times before returning to state 00. In other words, the minimum free distance is the weight of a code's minimum-weight error event. The code used in Example 7.2 has a minimum free distance of 5, and the error event shown in Fig. 7-22 is a minimum-weight event having a weight of 5. The branch from level 0 state 00 to level 1 state 10 has a weight of 2; the branch from level 1 state 10 to level 2 state 01 has a weight of 1; the branch from level 2 state 01 to level 3 state 00 has a weight of 2. Another way to look at this is that the encoder output corresponding to the error event path will contain exactly five 1's. At level 3 state 00, the incorrect path will be selected as the survivor if the received bit sequence contains 1's in more than 3 of the same locations in which the error-path encoded sequence contains 1's. (Remember, this analysis assumes that the all-zero codeword is being transmitted—thus each 1 in the received sequence represents a bit error.) In general, an incorrect path will be selected as the survivor if the received bit sequence contains 1's in more than half of the positions in which the error-path sequence contains 1's. Thus, a larger minimum free distance means that the decoder will be able to correct a larger number of bit errors in the received sequence.

A large value for d_{free} is a desirable characteristic for a "good" convolutional code. Unfortunately, there is no straightforward way to design convolutional codes for specific values of d_{free}. Most of the "good" codes have been found by computer search—*i.e.*, by exhaustively stepping through all possible polynomial sets and evaluating d_{free} for each code so generated.

7.2.2 Weight Distribution

Consider two hypothetical codes. Each code has $d_{\text{free}} = 5$. Code #1 has only one codeword of weight 5, and code #2 has 3 codewords of weight 5.

7.2.3 Information Sequence Weight

In Fig. 7-22, the complete error event spans 3 transitions, and the corresponding decoder output contains 1 bit error. In Fig. 7-24, the complete error event spans 6 transitions, and the corresponding decoder output contains 3 bit errors. Longer error events seem to cause more bit errors than shorter error events.

7.3 Viterbi Decoding with Soft Decisions

The Basic Idea

A major strength of the Viterbi decoder is its ability to readily make use of *soft decisions*. The advantage of soft decisions is that they contain some indication of how "good" the received signal was and hence how "confident" the receiver is regarding the decisions that have been made. A decoder can take advantage of this information by attaching greater importance to "high-confidence" decisions and less importance to "low-confidence" decisions. Loosely speaking, this varying importance is manifested as a greater reluctance on the part of the decoder to change high-confidence decisions and a greater willingness to change low-confidence decisions. This section takes a detailed look at an example use of soft decisions in a Viterbi decoder.

Consider the case of a hypothetical baseband channel in which a pulse of –1 volt is transmitted for one time unit to signify a bit value of 1, and a pulse of +1 volt is transmitted for one time unit to signify a bit value of 0. A *hard decision* receiver interprets received positive voltages as being bit values of 0, and received negative voltages as being bit values of 1. Intuitively, it is highly probable that a received voltage of –0.9 volt or –1.1 volts is caused by a transmitted voltage of –1.0 volt with a small amount of corruption introduced by the channel. Likewise, it is highly probable that a received voltage of +0.9 is caused by a transmitted voltage of +1.0 with a small amount of corruption introduced by the channel. We cannot be so confident about a received signal of –0.1 volt. Was the signal transmitted as –1.0 volt and subjected to 0.9 volt of channel corruption, or was it transmitted as +1.0 volt and subjected to 1.1 volts of channel corruption? Intuitively, we sense that voltages with magnitudes near 1.0 are more likely to have the correct sign than are voltages with magnitudes near zero. Unlike a hard decision receiver which, for this bipolar baseband signal, makes 1/0 decisions based on the sign of the received signal; a *soft decision* receiver makes use of the signal magnitude information as well.

Assume that in some particular channel the probability of a bit error is 0.1, or in other words, the probability of the received signal having the correct sign is 0.9. This means that if +1 volt is transmitted, the received voltage will be positive with a probability of 0.9. The distribution of received voltage corresponding to this situation is sketched in Fig. 7-25. Assume that in this channel we also know that if +1 volt is transmitted, the received voltage will be less than –0.5 V with probability 0.025, between –0.5 and 0 V with probability 0.075, between 0 V and +0.5 V with probability 0.3 and greater than +0.5 V with probability 0.6 (see Fig. 7-26). Likewise, if –1.0 volt is transmitted, then the received voltage will be greater than +0.5 V with probability 0.025, between +0.5 and 0 V with probability 0.075, between 0 and –0.5 V with probability 0.3, and less than –0.5 V with probability 0.6. This situation can be characterized as the *discrete memoryless channel (DMC)* depicted in Fig. 7-27.

We can exploit our knowledge about this channel to design a soft-decision receiver. If the received signal is greater than +0.5 V, the receiver can make a 0 decision with high confidence. We will designate this as the 0_H decision. If the received signal is between 0 and +0.5 V, the receiver can make a 0 decision with somewhat lower confidence. We will designate this as the 0_L decision. If the received signal is less than –0.5 V, the receiver can make a 1 decision with high confidence. We will designate this as the 1_H deci-

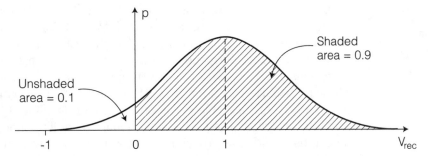

■ 7-25 *Distribution of received signal voltage caused by additive noise corruption of a transmitted voltage of +1.*

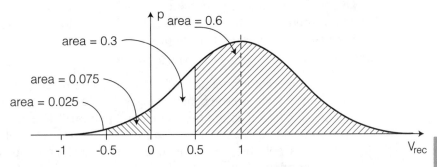

■ 7-26 *Distribution of received signal voltage (conditioned on +1 transmitted), divided into four regions corresponding to the four soft decisions 0_H, 0_L, 1_L, 1_H.*

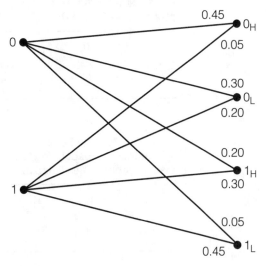

■ 7-27 *Binary to quaternary DMC.*

sion. If the received signal is between 0 and –0.5 V, the receiver can make a 1 decision with somewhat lower confidence. We will designate this as the 1_L decision. The transition probabilities shown in Fig. 7-27 correspond to the conditional probabilities of each decision (0_H, 0_L, 1_H, or 1_L) given the transmitted bit (0 or 1). These probabilities are summarized in Table 7-1. (The entry in column 1 row 1 of this table is the probability of deciding 0_H given that a 0 was transmitted. This conditional probability should not be confused with the probability that a 0 was transmitted given that the receiver decision is 0_H.

In working with soft decisions in a decoding trellis, we will be interested in the probability of **sequences** of transitions. The probability for a sequence of transitions would normally be computed as the product of the probabilities for the individual transitions. Because logarithms exhibit the property that

$$\log (a \bullet b) = \log a + \log b,$$

it is much more convenient (and numerically more sane) to work with metrics that are the common logarithms of the various probabilities. Instead of computing compound **probabilities** by multiplying individual probabilities (and obtaining values like 1.29×10^{-11}), it is possible, for logarithmic metrics, to compute compound **metrics** by adding the individual metrics (and obtaining values like –10.89). Table 7-2 lists the bit metrics obtained by taking the common logarithm of the probabilities in Table 7-1.

■ **Table 7-1 Transition probabilities for the DMC of Fig. 7-27.**

	0_H	0_L	1_L	1_H
0	0.45	0.3	0.2	0.05
1	0.05	0.2	0.3	0.45

■ **Table 7-2 Bit metrics corresponding to the transition probabilities listed in Table 7-1.**

	0_H	0_L	1_L	1_H
0	–0.35	–0.52	–0.7	–1.3
1	–1.3	–0.7	–0.52	–0.35

The five-error case represented by Figs. 7-18 and 7-19 had a hard-decision received sequence of 10 10 11 10 11 01 11 11 11. Assume that given the ability to make soft decisions, the decision sequence becomes 1_H0_L 1_H0_H 1_L1_L 1_H0_H 1_H1_H 0_L1_L 1_H1_H 1_H1_L 1_H1_H. Let's begin with the first received symbol 1_H 0_L and step through the process of generating the branch metrics for the first level of the decoding trellis:

1. The transmit symbol associated with the upper branch in the first level is 00. (We know that 11 was transmitted—not 00. However, the decoder does not "know" this, and in this step, the decoder is assessing the "reasonableness" of the decision pair 1_H0_L **if** 00 had been transmitted.)
   A. In Table 7-2, find the metric corresponding to the first transmit bit (0) and the first receiver decision (1_H). The value of this metric is −1.3.
   B. In Table 7-2, find the metric corresponding to the second transmit bit (0) and the second receiver decision (0_L). The value of this metric is −0.52.
   C. Add the two bit metrics to obtain the branch metric:
   $$(-1.3) + (-0.52) = -1.82$$

2. The transmit symbol associated with the lower branch in the first level is 11.
   A. In Table 7-2, find the metric corresponding to the first transmit bit (1) and the first receiver decision (1_H). The value of this metric is −0.35.
   B. In Table 7-2, find the metric corresponding to the second transmit bit (1) and the second receiver decision (0_L). The value of this metric is −0.7.
   C. Add the two bit metrics to obtain the branch metric:
   $$(-0.35) + (-0.7) = -1.05$$

The process outlined above is depicted in Fig. 7-28. If a similar process is followed for all 9 pairs of receiver decisions, we obtain the complete decoding trellis, which is shown with cumulative path metrics in Fig. 7-29. The path with the largest metric is highlighted. This is the correct path and it will result in the correct decoding of the transmitted sequence.[2] Using soft decisions has allowed us to correctly decode the five-error case, which we incorrectly decoded in Section 7.2 when limited to hard decisions.

---

[2] Probability increases with increasing "goodness" of the decisions, so the **largest** path metric is associated with the best set of decisions. This is in contrast to the Hamming-distance metric of Sections 7.1 and 7.2 in which distance decreases with increasing "goodness" of the decisions, causing the **smallest** path metric to be associated with the best set of decisions.

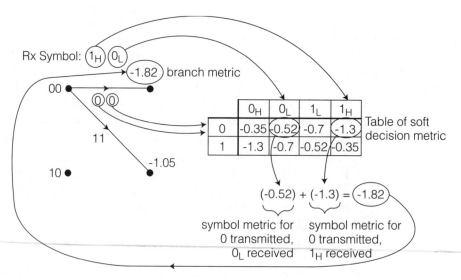

■ **7-28** *Using transmitted encoder outputs and receiver symbol decisions to generate branch metrics for the decoding trellis.*

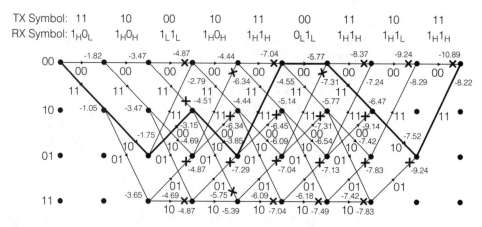

■ **7-29** *Decoding trellis showing use of real-valued path metrics for soft-decision decoding.*

7.4 Practical Issues

Except for the first few levels, each level in the decoding trellis will contain 2^K different states, with a different surviving path associated with each state. For some time, the conventional wisdom has held that Viterbi decoders are impractical for codes having constraint lengths longer than about 7 or 8. However, times change, and in the present era of ever-increasing hardware speed and ever-decreasing memory cost, this limit should be viewed as somewhat

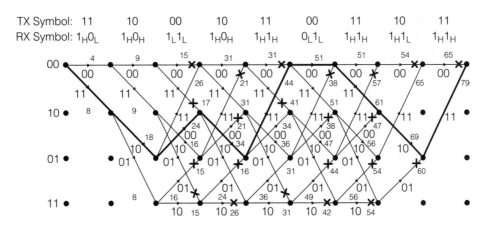

TX Symbol: 11   10   00   10   11   00   11   10   11
RX Symbol: 1_H0_L 1_H0_H 1_L1_L 1_H0_H 1_H1_H 0_L1_L 1_H1_H 1_H1_L 1_H1_H

■ **7-30** *Decoding trellis showing use of integer-valued path metrics for soft-decision decoding.*

flexible. For certain applications, the preferred approach might turn out to be Viterbi decoding of a convolutional code with a constraint length of 9 or 10.

7.4.1 Decoding Depth

For an information sequence length of kL, each path to be stored will ultimately consist of kL bits. For large L, the amount of storage required becomes impractical. In practical Viterbi decoder designs, only the $k\tau$ most recent bits of each path are stored. As each new level is processed, the decoder makes room for the new branches by making irrevocable bit decisions for the oldest level in storage. Decoding modified in this way is no longer a maximum likelihood decoder, but if τ is large enough, the performance will be very close to maximum likelihood.

Sequential Decoding

8

The computational complexity of a Viterbi decoder increases exponentially with the constraint length of the code being decoded. This complexity becomes impractical for constraint lengths larger than about 7 or 8. An alternate to Viterbi decoding, *sequential decoding*, has a complexity that does not depend upon constraint length. This property makes sequential decoding suitable for use with codes having long constraint lengths. The two most well known types of sequential decoding—*stack decoding* and *Fano decoding*—are presented in this chapter.

8.1 Stack Decoding Algorithm

The Basic Idea

Stack decoding proceeds by building a number of partial paths through the code tree, and placing these paths on a stack. New partial paths are created by extending the "best" partial path already on the stack, with "best" being determined via a metric that has been optimized for making comparisons between partial paths of unequal lengths. For an m-ary code, the best path is extended in m different ways by appending in turn each of the symbols 0, 1, ...m to construct m new partial paths each one symbol longer than the "best" path upon which they have been built. The new "best" path is determined, and the process of extension is repeated. After a number of iterations, the "best" path will comprise a complete path through the trellis. The sequence of encoder input bits corresponding to this complete path is issued as the decoded output sequence.

Assume that we have a state table, state diagram, or other equivalent description of a convolutional encoder's operation. We also have a sequence of rules for using this information to generate a set of paths through a decoding tree. In general, the paths in this set will have different lengths, and each path will have associated with it a metric that roughly corresponds to the relative likelihood of that particular path being the correct path. As the algorithm progresses, a list or *stack* of the various paths is maintained in order according to metric, with the path having the "best" metric placed at the top of the stack.[1] During each iteration of the algorithm, the path at the top of the stack is extended by one level and the resulting new paths are inserted into the stack at the appropriate locations according to their metrics. When a complete path through the tree appears at the top of the stack, the algorithm stops and this path is deemed to be the correct path. The sequence of inputs corresponding to this path is issued as the decoded output.

Example 8.1 Once again, consider the "canonical" example. The state diagram is reproduced here as Fig. 8-1. Each codeword is produced from 7 information bits plus a 2-bit tail of zeros added to flush out the encoder. The 7-bit ASCII code for Q is 1010001, which is encoded as the 18-bit codeword

$$11 \ 10 \ 00 \ 10 \ 11 \ 00 \ 11 \ 10 \ 11$$

The signal is transmitted over a binary symmetric channel (BSC). Three bit errors occur, and the received codeword is

$$11 \ 00 \ 00 \ 00 \ 11 \ 00 \ 01 \ 10 \ 11$$

Use the Hamming metric to perform stack decoding on this codeword.

Solution (Cycle #1) Consulting the state diagram of Fig. 8-1, we determine:

☐ the encoder will remain in state 0 if the input bit is 0

☐ the encoder will transition from state 0 to state 2 if the input is 1

If the input is 0, the first encoder output symbol is 00. The first received symbol is 11, so we take the Hamming distance between 00 and 11 to obtain 2 as the branch metric. Similarly, if the first input is 1, the encoder output symbol is 11, and the branch metric is 0. The resulting stack contents after processing of the first received symbol are shown in Table 8-1. The corresponding partial tree is shown in Fig. 8-2.

---

1 For most metrics, the "best" metric will be the largest. However, for a few metrics, the "best" will be the smallest.

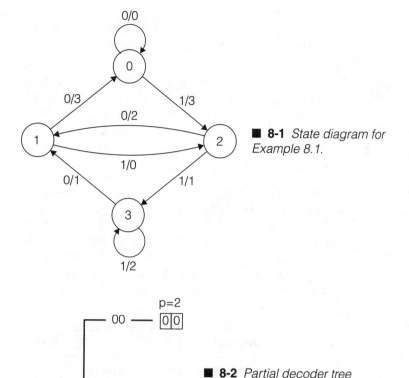

8-1 *State diagram for Example 8.1.*

p=2

00 — [0|0]

8-2 *Partial decoder tree for Example 8.1 after first decoding cycle.*

[0|0]

11 — [1|0]

p=0

■ Table 8-1 Stack contents for Example 8.1 after the first processing cycle.

Inputs	Encoder states	Candidate outputs	Path metric
1	0,2	1 1	0
0	0,0	0 0	2

(Cycle #2) Working with the top entry in Table 8-1, we find that the final encoder state is 2. Therefore, we next consider the consequences of the second encoder input bit assuming that the encoder was in state 2 prior to the arrival of the input.

☐ If this second input is 0, the encoder will transition from state 2 to state 1 and issue 10 as output. The second received symbol is 00, so we take the Hamming distance between 10 and 00 to obtain 1 as the branch metric. Adding this branch metric to the top path metric in Table 8-1 results in the new path metric of 1.

□ If the second encoder input is 1, the encoder will transition from state 2 to state 3 and issue 01 as output. The second received symbol is 00, so we take the Hamming distance between 01 and 00 to obtain 1 as the branch metric. Adding this branch metric to the top path metric in Table 8-1 results in the new path metric of 1.

The two new partial paths created in cycle 2 each have a metric of 1, which is smaller than the other metric remaining in the table. Therefore, either of the two new metrics could be placed at the top of the stack with the other being placed second. We will arbitrarily place the path corresponding to the input sequence 10 at the top of the stack to obtain the stack contents shown in Table 8-2. The corresponding partial tree is shown in Fig. 8-3.

■ Table 8-2 Stack contents for
Example 8.1 after the second processing cycle.

Inputs	Encoder states	Candidate outputs	Path metric
1,0	0,2,1	11,10	1
1,1	0,2,3	11,01	1
0	0,0	00	2

(Cycle #3) Starting with the top entry in Table 8-3, we next consider the possibilities associated with the third encoder input bit assuming that the encoder is in state 1 prior to the arrival of the input:

□ If the third input is 0, the encoder will transition from state 1 to state 0 and issue 11 as output. The third received symbol is 00, so the branch metric is dist(11,00) = 2. Adding this branch metric to the top path metric in Table 8-2 results in the new path metric of 3. The stack entry for this new partial path is placed at the bottom of the new stack because the metric is larger than any of the other metrics in the stack.

□ If the third input is 1, the encoder will transition from state 1 to state 2 and issue 00 as output. The branch metric is (00,00) = 0, and the partial path metric is 1 + 0=1. The stack entry for this new partial path is placed at the top of the new stack because the metric is smaller than any of the other metrics in the stack.

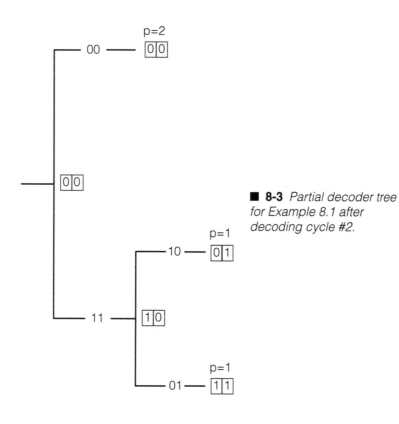

p=2
00 — 0 0

0 0

■ **8-3** *Partial decoder tree for Example 8.1 after decoding cycle #2.*

p=1
10 — 0 1

11 — 1 0

p=1
01 — 1 1

159

The resulting stack contents are shown in Fig. 8-3. The corresponding partial tree is shown in Fig. 8-4.

■ **Table 8-3 Stack contents for Example 8.1 after the third processing cycle.**

Inputs	Encoder states	Candidate outputs	Path metric
1,0,1	0,2,1,2	11,10,00	1
1,1	0,2,3	11,01	1
0	0,0	00	2
1,0,0	0,2,1,0	11,10,11	3

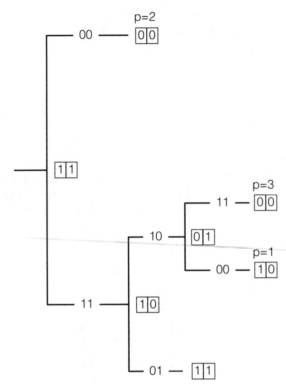

■ **8-4** *Partial decoder tree for Example 8.1 after decoding cycle #3.*

■ **Table 8-4**
Streamlined version of Table 8.3.

Inputs	Final Encoder states	Path metric
101	2	1
11	3	1
0	0	2
100	0	3

After a few cycles, it quickly becomes unwieldy to keep recording the stack contents in the format of Table 8-3. All that is really needed for each entry is the input sequence, the final encoder state resulting from this sequence, and the metric for the corresponding partial path. We can rewrite Table 8-3 in a more streamlined format as shown in Table 8-4. Continuing the process for 17 more cycles produces the sequence of stack contents as shown in Tables A-4 through A-20, which are in Appendix A. In each of these tables, the two most recently inserted entries are marked with an asterisk. The entries that eventually become the correctly decoded output are marked with a dagger (†). The last 3 processing cycles are different than the others. Because of the "tail" of 0-valued inputs used to flush the encoder after the entire 7-bit input has been entered, only one possibility needs to be considered for these last two cycles. After the processing cycle (20), the path at the top of the stack is complete with 9 inputs (7 for data, 2 for flushing), so the process stops and the sequence 1010001 is issued as the decoder output. Figure 8-5 shows the complete path corresponding to the top of the stack as well as all the other partial paths that were created along the way.

One of the problems frequently encountered in using the stack algorithm is the large number of processing cycles that may be required to decode a received frame. In Example 8.1, 20 processing cycles were required to decode a 9-symbol frame. The number of required cycles can be reduced if a different metric is used in place of the Hamming metric. To understand why this is so, look at the tables of Appendix A. In many of the tables, the top path is not the partial path, which eventually becomes the correct complete path. In Fig. 8-6, the stack position of the correct partial path is plotted as a function of cycle number. Why does the correct path spend so much time out of the top position? At cycle 8, the metric of the correct path increases to 3.0 while the metrics of four incorrect paths remain at 2.0. These incorrect paths have lengths of 4, 3, 3, and 1, but the correct path in stack position 6 has a length of 7. It takes five additional cycles for the metrics of the incorrect paths to degrade to the point where they all move to stack positions below the correct path. The problem seems to revolve around comparing short paths with small metrics against (significantly) longer paths with (slightly) larger metrics. The results of such comparisons often push the correct path out of the top slot because the Hamming metric does not penalize short paths or reward long paths. The Fano metric presented in the following section is one approach for obtaining better comparisons between paths of unequal length.

8.1.1 Fano Metric

Recognizing that sequential decoding would benefit from a metric that takes path length into account, Fano[2] devised a metric that

2 R. M. Fano, "A Heuristic Discussion of Probabilistic Decoding," *IEEE Trans. Inf. Theory*, IT-9, pp. 64-74, April 1963.

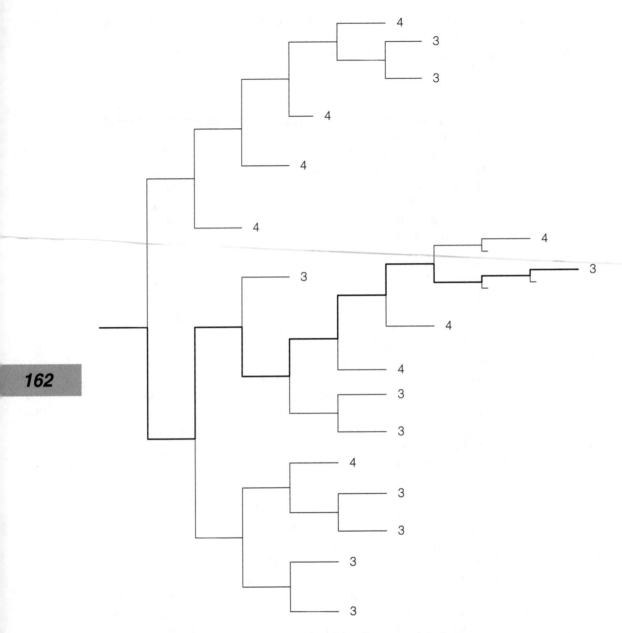

■ **8-5** *Decoder tree for Example 8.1 after decoding completed.*

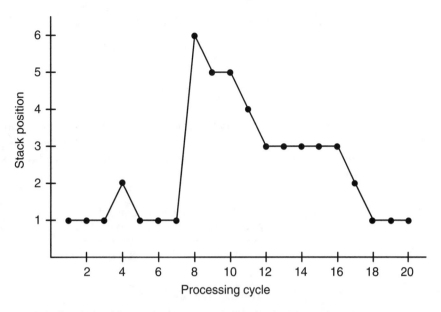

■ **8-6** *Stack positions of correct partial paths for Example 8.1.*

now bears his name. The Fano metric $M(r|v)$ for received symbol r and candidate transmit symbol v is given by

$$M(r|v) = \log_2 \frac{P(r|v)}{P(r)} - R$$

where $P(r)$ is the channel output symbol probability, $P(r|v)$ is the conditional probability of receiving symbol r given that symbol v was transmitted, and R is the code rate. When the Fano metric is used, the metric of the correct path tends to increase at a moderate rate, and the metrics of incorrect paths decrease relatively quickly. Fano obtained this metric via a heuristic approach, but a rigorous treatment was subsequently published by Massey.[3]

Example 8.2 Repeat Example 8.1 using the Fano metric instead of the Hamming metric. Assume that the channel has a bit error rate of 0.05.

Solution In a BSC with a BER of 0.05, the conditional probabilities for the 4 possible combinations of input and output are:

Prob(receiving 0 given 0 transmitted) = 0.95
Prob(receiving 0 given 1 transmitted) = 0.05
Prob(receiving 1 given 0 transmitted) = 0.05
Prob(receiving 1 given 1 transmitted) = 0.95

---

3 J. L. Massey, "Variable-Length Codes and the Fano Metric," *IEEE Trans. Inf. Theory,* IT-18, pp. 196-198, January 1972.

We compute the Fano metrics as follows:

$$3.322 \left(\log \frac{0.95}{0.5} \right) - \frac{1}{2} = 0.426$$

$$3.322 \left(\log \frac{0.05}{0.5} \right) - \frac{1}{2} = -3.822$$

The conditional probabilities and resulting bit metrics are summarized in Tables 8-5 and 8-6. These bit metrics can be added together in the appropriate combinations to form the symbol metrics listed in Table 8-7.

■ **Table 8-5 Conditional bit probabilities $P(r|v)$ for Example 8.2.**

		Received bit, r	
		0	**1**
Transmitted	0	0.95	0.05
bit, v	1	0.05	0.95

■ **Table 8-6 Bit metrics $M(r|v)$ for Example 8.2.**

		Received bit, r	
		0	**1**
Transmitted	0	0.426	−3.822
bit, v	1	−3.822	0.426

■ **Table 8-7 Symbol metrics for Example 8.2.**

		Transmitted symbol			
		00	**01**	**10**	**11**
	00	0.852	−3.396	−3.396	−7.644
Received	01	−3.396	0.852	−7.644	−3.396
symbol	10	−3.396	−7.644	0.852	−3.396
	11	−7.644	−3.396	−3.396	0.852

(Cycle #1) As usual, we assume that the encoder begins in state 0. Consulting the state diagram of Fig. 8-1, we determine:

☐ the encoder will remain in state 0 if the input bit is 0

☐ the encoder will transition from state 0 to state 2 if the input is 1

Parent Worksheet

Swim Classes	Session 1 Tuesday and Thursdays Sept 7 – Sept 30	Session 2 Tue/Wed 10 weeks Thur/Fri/Sat 8 weeks Oct 2 – Dec 15 No Monday Classes	Session 3 All Classes 10 weeks Jan 4 – March 19 No Monday Classes	Session 4 All Classes 8 weeks March 26 – May 27
Water Babies	NA	Sat 8:45-9:15, 12:30-1:00		
Parent Toddler	NA			
Preschool	NA	Wed 10:00-10:30, Sat 9:15-9:45, 12:30-1:00		
Beginner	11:00-11:30, 3:30-4:00,	Mon 3:30-4:00, Tue 11:00-11:30, 3:30-4:00, Wed 10:30-11:30, Thur 11:00-11:30, 1:00-1:30, 3:30-4:00, Fri 11:00-11:30, Sat 9:45-10:15, 10:15-10:45, 11:00-11:30		
Adv Beginner	1:30-12:00, 3:30-4:00,	Mon 3:30-4:00, Tue 11:30-12:00, 1:30-2:00, 4:00-4:30, 3:30-4:00, Wed 11:30-12:00, Thur 11:30-12:00, Fri 11:30-12, Sat 10:15-10:45, 10:30-11:00, 11:30-12:00, 12:00-12:30		
Intermediate	10:00-10:30, 4:00-4:30,	Mon 4:00-4:30, Tue 10:00-10:30, 2:00-2:30, 4:00-4:30, 4:30-5:00, Wed 11:00-11:30, Thur 10:00-10:30, 11:30-12:00, Fri 10:00-10:30, Sat 10:15-10:45, 12:00-12:30		
Swimmer	10:30-11:00, 4:45-5:15,	Mon 4:45-5:15, Tue 10:30-11:00, 4:45-5:15, Thur 10:30-11:00, 4:30-5:00, Fri 10:30-11:00, Sat 9:45-10:15, 11:30-12:00		
School Age				
Level 1	4:00-4:45	Mon 4:00-4:45, Tue 4:00-4:45, Thur 3:30-4:15, Sat 11:30-12:15		
Level 2	4:30-5:15	Mon 4:30-5:15, Tue 4:30-5:15, Thur 4:15-5:00,		
Level 3	5:15-6:00	Mon 5:15-6:00, Tue 5:15-6:00, Thur 5:00-5:45, Sat9:45-10:30, 12:15-1:00		
Level 4	5:15-6:00	Mon 5:15-6:00, Tue 5:15-6:00, Thur 5:00-5:45, Sat 9:30-10:15, 11:30-12:15		
Level 5 and 6	NA	Mon 5:15-6:00, Tue 5:15-6:00, Sat 12:15-1:00		
Tennis and Swim Combo	NA	Thur 5:00-5:45, Sat 8:45-9:30, Thur 4:00-6:00, Sat 9:30-11:30		

Swim Lesson Rates For Session 3

Drop off Registration for the lottery is Mon, Dec 13 - Wed Dec 15
Classes start the week of January 4, 2005.

	Members	Public
30 minute classes Waterbabies, Parent Toddler Beginners, Adv. Beginners Intermeidate Swimmer	$62.00	$101.00
45 minute classes Beginners, Adv. Beginners Intermeidate Swimmer	$83.00	$120.00

2004 – 2005 Registration Form

(Please fill out a separate form for each child and each program)
(To receive the member rate the child must be a member)

Child's Name _____ Age _____

E mail _____ Date of Birth ____/____/____

Phone # _____ Member #_____

Address: _____

_____

Parent's Name_____

Tennis   Racquetball   Session   I   II   III   IV

    Swimming   Fall   Winter   Spring

	Class Level	Class Time	Class Day
Choice 1			
Choice 2			
Choice 3			

Class Cost: $_____

(Payment must accompany registration form)

-----------------------------------Office use only-------------------------------------

Payment Method

❏ Credit Card

❏ House Charge

❏ Cash

❏ Check

Charge Codes

624   Tennis and Swim
705   Tennis
730   Racquetball
800   Swimming

Chit # _____

2004

(Please
(To

Child's Name _____

E mail _____

Phone # _____

Address: _____

Parent's Name_____

Tennis   Racquetball

    Swimming

	Class Level	Cl
Choice 1		
Choice 2		
Choice 3		

Cla

(Payment must accompany registration form)

-----------------------------------Office use only-------------------

Payment Method

❏ Credit Card

❏ House Charge

❏ Cash

❏ Check

Charge Code

624   Tennis a
705   Tennis
730   Racquetbal
800   Swimming

Chit # _____

Level 1 - Lev

The Classes in the Winter Session mee
There are no Monday Classes in the Winter.
Those who participate in the lottery will be notified by Wednesday.
December 22, 2004 of their enrollment status.

If the input is 0, the first encoder output symbol is 00. Because the first received symbol is 11, the metric for this branch (–7.644) is given by the entry in row 11 column 00 of Table 8-7. Similarly, if the input is 1, the first encoder output is 11. The metric for this branch (0.852) is given by the entry in row 11, column 11 of Table 8-8. The resulting stack contents after processing of the first received symbol are show in Table 8-7. The corresponding partial tree is shown in Fig. 8-7.

■ Table 8-8 Stack contents for
Example 8.2 after first processing cycle.

Inputs	Final encoder state	Path metric
†1	2	0.852
0	0	–7.644

(Cycle #2) Working with the top entry in Table 8-8, we find that the final encoder state is 2. Therefore, we next consider the consequences of the second encoder input bit assuming that the encoder was in state 2 prior to the arrival of the input.

☐ If this second input is 0, the encoder will transition from state 2 to state 1 and issue 10 as output. Since the second received symbol is 00, we find the branch metric of –3.396 from row 10 column 00 of Table 8-7. Adding this branch metric to the top path metric in Table 8-8, results in the new path metric of –2.544.

☐ If the second encoder input is 1, the encoder will transition from state 2 to state 3 and issue 01 as output. The second

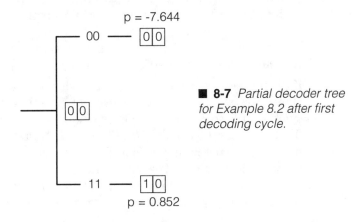

■ 8-7 *Partial decoder tree for Example 8.2 after first decoding cycle.*

received symbol is 00, so we find the corresponding branch metric of –3.396 from row 01 column 00 of Table 8-7. Adding this branch metric to the top path metric in Table 8-8 results in the new path metric of –2.544. Because both new entries have equal path metrics (which are larger than the other entry in the stack) the choice of which entry to place at the top of the stack is arbitrary. When two new entries have equal metrics, we will follow the convention of placing the entry corresponding to a 0 input ahead of the entry corresponding to a 1 input.

The resulting stack contents are shown in Table 8-9, and the corresponding partial tree is shown in Fig. 8-8.

■ **Table 8-9 Stack contents for Example 8.2 after second processing cycle.**

Inputs	Final encoder state	Path metric
†*10	1	–2.544
*11	3	–2.544
0	0	–7.644

(Cycle #3) Starting with the top entry in Table 8-9, we next consider the possibilities associated with the third encoder input bit, assuming that the encoder is in state 1 prior to the arrival of the input:

☐ If this third input is 0, the encoder will transition from state 1 to state 0 and issue 11 as output. Because the third received symbol is 00, we find the branch metric of –7.644 from row 00 column 11 of Table 8-7. Adding this branch metric to the top path metric in Table 8-9 results in the new path metric of –10.188. The stack entry for this new partial path is placed at the bottom of the new stack because the metric is smaller than any of the other metrics in the stack.

☐ If the third encoder input is 1, the encoder will transition from state 1 to state 2 and issue 00 as output. The third received symbol is 00, so we find the corresponding branch metric of 0.852 from row 00 column 00 of Table 8-7. Adding this branch metric to the top path metric in Table 8-9 results in a new path metric of –1.692. The stack entry for this new partial path is placed at the top of the new stack because the metric is larger than any of the other metrics remaining in the stack.

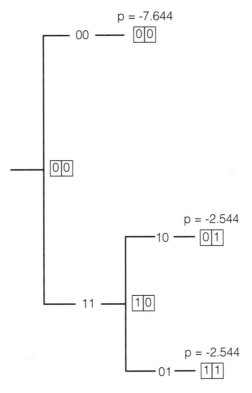

p = -7.644

00 —— 0|0

0|0

p = -2.544

10 —— 0|1

11 —— 1|0

p = -2.544

01 —— 1|1

■ **8-8** *Partial decoder tree for Example 8.2 after decoding cycle #2.*

The resulting stack contents are shown in Table 8-10. The corresponding partial tree is shown in Fig. 8-9.

■ **Table 8-10 Stack contents for Example 8.2 after third processing cycle.**

Inputs	Final encoder state	Path metric
†*101	2	−1.692
11	3	−2.544
0	0	−7.644
*100	0	−10.188

Continuing the process for 12 more cycles produces the sequence of stack contents as shown in Tables B-4 through B-15, which can

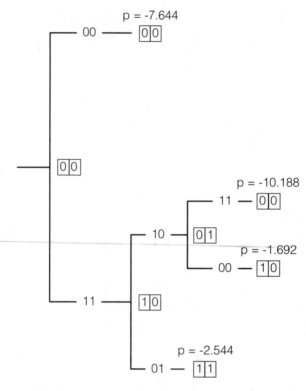

■ 8-9 *Partial decoder tree for Example 8.2 after decoding cycle #3.*

be found in Appendix B. After cycle #15, the path on the top of the stack is complete with 9 inputs (7 for data, 2 for flushing), so the process stops and the sequence 1010001 is issued as the decoder output. Figure 8-10 shows the complete path corresponding to the top of the stack as well as all the other partial paths that were created along the way.

Using the Fano metric, Example 8.2 completes decoding in 15 cycles. This is a significant improvement over the 20 cycles required when using the Hamming metric in Example 8.1. Figure 8-11 compares the correct paths' stack positions for Examples 8.1 and 8.2.

8.2 Software for Stack Decoding

The stack decoding algorithm illustrated by Examples 8.1 and 8.2 is more formally stated in Recipe 8.1.

Recipe 8.1 [Stack decoding.]

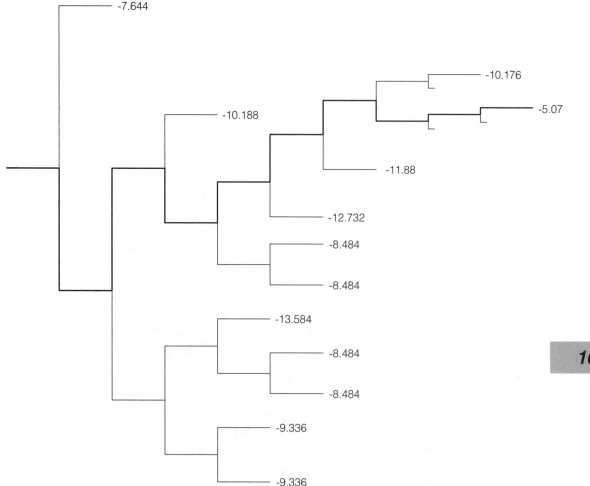

■ **8-10** *Decoder tree for Example 8.2 after decoding completed.*

1. Obtain the definition of the particular code to be decoded. This definition will be most useful in the form of either a state diagram or a state table.
2. Based on the channel characteristics, create a table of bit metrics.
3. Based on the details of the symbol structure, use the bit metrics from step #2 to create a table of symbol metrics.
4. Initialize the stack with a single entry—null input sequence, final state of 0, and path metric of 0.

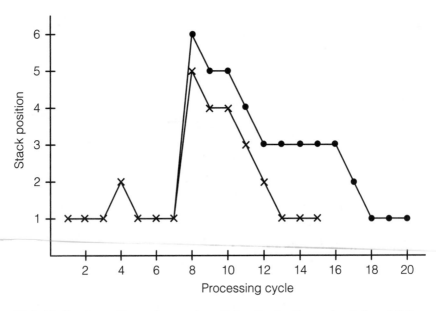

■ **8-11** *Stack positions of correct partial paths for Examples 8.1 and 8.2.*

5. Examine the top entry on the stack. Let L_T denote the number of symbols in the top entry's input sequence. Let S_T denote the final state indicated in the top entry. For each one of the b possible values of an input symbol, create a new input sequence of $L_T + 1$ symbols:
   A. Consult the state table or state diagram to determine the next state that will result from each of the possible b inputs given that the encoder starts in state S_T.
   B. Determine the encoder output corresponding to each of these state transitions.
   C. Use the encoder output in conjunction with the received symbol numbered $L_T + 1$ to determine the branch metric for each of the transition branches. Add each branch metric to the path metric of the "top" path to determine the new path metric for the corresponding new path.

6. Remove the top path from the stack.

7. Insert each of the b new paths created in step #5 into the appropriate locations within the ordered stack of partial paths.

8. If the path now occupying the top stack location spans the complete code word, this path represents the decoder's estimate of the encoder path most likely to have been transmitted. Otherwise jump back to step #5 and continue building paths through the tree.

8.2.1 Implementing the Stack

If we are to implement the stack algorithm in software, one obvious requirement is the allocation of memory to hold the stack elements. Before this allocation can be implemented, we need to find answers to a number of questions. Should a fixed amount of memory be allocated, or should we design for dynamic allocation of stack space on an as-needed basis? What should the internal structure of a stack element be? How should the algorithm software interface to the stack elements?

Stack Elements

The stack tables of Appendices A and B have columns for partial sequences of input bits, final encoder states, and path metrics. It is a straightforward matter to store final encoder states as int values and to store path metrics as float values. The approach for storing the partial sequences of input bits is not so obvious. Although the manually generated tables of the appendices show all of the bits in each input sequence, this is not really necessary. In fact, throughout most of the algorithm, the top-level software never really needs to have "read" access to the partial input sequences. The top-level software only needs to: (1) Append new input symbols to the partial input sequence at the top of the stack, and (2) Check any partial input sequence to see if it defines a complete path through the encoder trellis. Neither of these operations requires that the top-level software be provided with the contents of a partial input sequence. Once the input sequence at the top of the stack does define a complete path through the trellis, then and only then, do we need a method for "reading" the contents of this sequence because these contents will be issued as the decoder output.

Stack Structure

One of the most crucial operations in Recipe 8.1 occurs in step 7, where each newly extended path must be inserted into the stack at the appropriate location based upon the value of its path metric. Exactly how this insertion is accomplished depends upon how the stack is actually implemented. It is inconvenient that the algorithm under discussion has come to be known as the *stack algorithm* and that consequently the tables of Appendices A and B are referred to as "stacks." In computer science literature, the term *stack* is used to refer to a single-port last-in-first-out (LIFO) buffer structure. The two usages of "stack" are related, but not identical. For the sake of clarity, we will begin referring to the stacks of Appendices A and B as *stack tables*, and to the stacks of computer science as *stack buffers*. The stack tables used for the stack decoding algorithm can be implemented in software using either *linked-lists* or stack buffers.

Linked-List Implementation of the Stack The concept behind a *linked-list* data structure is depicted in Fig. 8-12. The individual entries within the list need not be stored contiguously or even in order. As shown, entry_1, entry_2, and entry_3 are actually stored in the order entry_1, entry_3, entry_2. Furthermore, between entry_3 and entry_2, there is a section of memory that is used for something other than the linked list under discussion. The order of the list is maintained by linking the individual entries. The last location in entry_1 contains a link that points to address_C, which is the location where entry_2 begins. Likewise, the last location in entry_2 contains a link that points to address_B, which is the location where entry_3 begins.

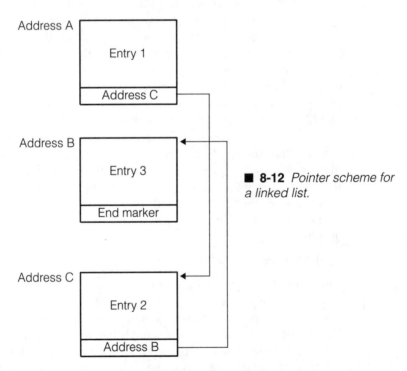

■ **8-12** *Pointer scheme for a linked list.*

Inserting a new entry into the middle of a linked list can be accomplished without having to move any entries to make room for the new entry. Assume that we wish to insert a new entry into the list between entry_1 and entry_2. As depicted in Fig. 8-13, the new entry is placed in memory beginning at address_D. The link in entry_1 is changed to point to address_D, and the link in new_entry is

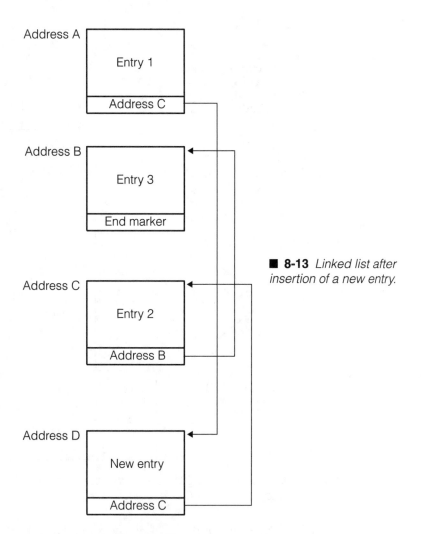

■ **8-13** *Linked list after insertion of a new entry.*

173

set to point to address_C. If a linked list is used to store the stack table entries for the stack decoding algorithm, the entries for newly extended partial paths can be inserted into the list without having to physically rearrange any of the other entries. Unfortunately, it is not quite this simple to find the appropriate location within the list in which to place a new entry. The decoding algorithm would have to traverse the list entries and examine their metric values and input sequence lengths according to the scheme defined by Recipe 8.2 and illustrated by the flow chart in Fig. 8-14.

Recipe 8.2 [Finding the insertion point for a new list entry.]

1. Access next entry in list. Let M denote the path metric of this entry, and let L denote the input sequence length for this entry.

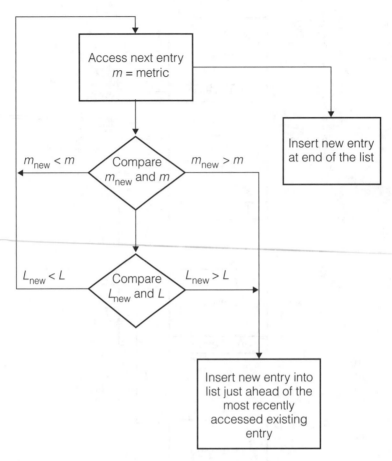

■ **8-14** *Flow chart for finding the new-entry insertion point in a linked-list implementation of the stack decoding algorithm.*

2. Compare M_{new} and M. If $M_{new} > M$, go to step #5. If $M_{new} < M$, go to step #4.

3. Compare L_{new} and L. If $L_{new} \geq L$, go to step #5.

4. Check pointer to next entry in list. If not end of list, go to step #1. If at end of list, insert new entry at end of list and **stop**.

5. Insert new entry into list just ahead of the most recently accessed existing entry and **stop.**

Stack Implementation of the Stack As mentioned previously, a stack is a last-in-first-out (LIFO) buffer. A stack gets its name from the fact that it is conceptually viewed as a stack of entries (Fig. 8-15A). Just like a stack of dishes, new items are added to or removed from the stack **only at the top**. Furthermore, items are added to a stack one at a time. The operation of adding an item to the top of

entry_4
entry_3
entry_2
entry_1

entry_B
entry_A
entry_4
entry_3
entry_2
entry_1

(A)   (B)

■ **8-15** *Operation of a last-in-first-out buffer or stack: (A) original stack, (B) stack after addition of entry _A and entry _B.*

a stack is often referred to as a *push* operation. The operation of removing an item from the top of a stack is often referred to as a *pop* operation. If we add two new entries—for example, entry_A and entry_B—to the stack depicted in Fig. 8-15A, the new stack configuration will be as shown in Fig. 8-15B. We first push entry_A onto the stack, then entry_B. Recipe 8-3 shows how one iteration of the stack algorithm can be performed in terms of push and pop operations. (This algorithm is implemented in the file stackalg.c contained on the disk accompanying this book.)

Recipe 8.3 [Stack implementation of one iteration for the stack decoding algorithm.]

1. **Pop** the top entry off of the main stack. Designate this entry as "Entry X." This entry consists of three parts: (1) a partial sequence of input bits, (2) the final state of the encoder after processing the partial sequence of input bits, and (3) a path metric for the encoder-trellis path corresponding to the given sequence of input bits.

2. Make a copy of Entry X to create Entry Y.

3. Extend Entry X by appending a 0 bit to the input sequence. Update the final encoder state and the path metric M_X accordingly.

4. If processing a "tail" symbol (*i.e.*, if the extended input sequence in Entry X is longer than the length of the code's information field), then redesignate Entry X as Entry S and go to step #16. Otherwise, continue.

5. Extend Entry Y by appending a 1 bit to the input sequence. Update the final encoder state and the path metric M_Y accordingly.

6. Compare metrics for Entry X and Entry Y. If $M_X \geq M_Y$ redesignate Entry X as "Entry L," and redesignate Entry Y as "Entry S." Otherwise (*i.e.*, if $M_X < M_Y$), redesignate Entry X as "Entry S," and redesignate Entry Y as "Entry L."

7. Loop while main stack is not empty:
   A. Get a new Entry X by **popping** next entry off main stack.
   B. Compare M_L and M_X. If $M_L > M_X$, go to step #11. If $M_L < M_X$, go to step #7D.
   C. Compare L_L and L_X. If $L_L \geq L_X$, go to step #11.
   D. **Push** Entry X onto auxiliary stack.

8. **Push** Entry S onto main stack.

9. **Push** Entry L onto main stack.

10. **Pop** an entry from the auxiliary stack and **push** it onto the main stack. Repeat until auxiliary stack is empty. Once the auxiliary stack is empty, this iteration is complete. The main stack contains the updated stack with Entry L and Entry S located at the bottom of the stack.

11. **Push** Entry L onto auxiliary stack.

12. Compare M_S and M_X. If $M_S < M_X$, **push** Entry X onto the auxiliary stack and go to step #16. Otherwise, continue.

13. **Push** Entry X onto main stack.

14. **Push** Entry S onto main stack.

15. **Pop** an entry from the auxiliary stack and **push** it onto the main stack. Repeat until auxiliary stack is empty. Once the auxiliary stack is empty, this iteration is complete. The main stack contains the updated stack with Entry L and Entry S in consecutive locations somewhere in the middle of the stack.

16. Loop while main stack is not empty:
   A. Get new Entry X by **popping** next entry off main stack.
   B. Compare M_S and M_X. If $M_S > M_X$, go to step #19. If $M_S < M_X$, go to step #16D.
   C. Compare L_S and L_X. If $L_L \geq L_X$, go to step #19.
   D. **Push** Entry X onto auxiliary stack.

17. **Push** Entry S onto main stack.

18. **Pop** an entry from the auxiliary stack and **push** it onto the main stack. Repeat until auxiliary stack is empty. Once the auxiliary stack is empty, this iteration is complete. The main stack contains the updated stack with Entry S located at the bottom of the stack and Entry L located elsewhere in the stack.

19. **Push** Entry X onto main stack.

20. **Push** Entry S onto main stack.

21. **Pop** an entry from the auxiliary stack and **push** it onto the main stack. Repeat until auxiliary stack is empty. Once the auxiliary stack is empty, this iteration is complete. The main stack contains the updated stack with Entry L and Entry S in nonconsecutive locations somewhere in the middle of the stack.

8.2.2 Received Symbol Buffering

Examples 8.1 and 8.2 are relatively "clean." The correctly decoded message is obtained in both cases. The need for received symbol buffering is better illustrated with a somewhat less benign example.

Example 8.3 Repeat Example 8.2 for the case of the following received sequence which contains 5 errors:

$$10 \;\; 10 \;\; 11 \;\; 10 \;\; 11 \;\; 01 \;\; 11 \;\; 11 \;\; 11$$

Solution The stack contents after each of 25 processing steps are listed in the tables of Appendix C. The output of the decoder (which is obtained as the first 7 bits of the top entry in Table 8D.25) is 0010001. This result contains a single bit error in the first position.

Table 8-11 lists the received symbol used by each iteration of the stack algorithm in Example 8.3. Unlike Example 8.2, a single path does not continue to grow while remaining at the top of the stack. The incumbent top path is replaced by a different path after steps 2, 8, 9, 11, 12, 16, 17, 19, 20, and 21. When iteration k of the algorithm puts a shorter partial path at the top of the stack, iteration $k + 1$ will need to go back and examine received symbols that were received earlier. As late as iteration 17 in Example 8.3, the algorithm needs to make use of received symbol number 3. Clearly, the received message symbols need to be buffered for possible use by later iterations of the algorithm. In our short 9-symbol message frame used in the examples, this buffer requirement is minor, but for longer frames, the buffering can become a burden. It might seem as though a buffer long enough to hold a complete frame would be sufficient, but it turns out that the actual requirement is much worse.

■ Table 8-11 Received symbols used for stack decoding of the 5-error case of Example 8.3.

Cycle	Receive symbol needed	Newest receive symbol available
1	1	1 (frame 1)
2	2	2 (frame 1)
3	2	3 (frame 1)
4	3	4 (frame 1)
5	4	5 (frame 1)
6	5	6 (frame 1)
7	6	7 (frame 1)
8	7	8 (frame 1)
9	5	9 (frame 1)
10	8	1 (frame 2)
11	9	2 (frame 2)
12	8	3 (frame 2)
13	3	4 (frame 2)
14	4	5 (frame 2)
15	5	6 (frame 2)
16	6	7 (frame 2)
17	3	8 (frame 2)
18	7	9 (frame 2)
19	8	1 (frame 3)
20	7	2 (frame 3)
21	6	3 (frame 3)
22	6	4 (frame 3)
23	7	5 (frame 3)
24	8	6 (frame 3)
25	9	7 (frame 3)

Assume that each iteration of the stack algorithm takes an amount of time equal to the duration of one received symbol. The third column of Table 8-11 reveals that by the time the algorithm completes iteration 25 in the decoding of frame #1, the buffer will need to be holding all of frame #1, all of frame #2, and the first seven symbols of frame #3. This example illustrates the problem, but it does not go far enough with regard to showing just how big the buffer really needs to be. For practical frame lengths, it turns out that even if the buffer is 100 times longer than the frame length, there is some nonzero probability that the buffer will fill up and be forced to discard a symbol that might still be needed by the decoder. If a symbol is needed for decoding after it has been pushed out of the buffer, the decoder will not be able to decode the frame. The resulting lost frame is called an *erasure*. Typical decoders are designed to have erasure probabilites of approximately 10^{-3}.

Erasures are not always as bad as they sound. An erasure is most likely to occur when the decoder is working on a noisy frame that would likely be decoded incorrectly anyway. In an ARQ scheme, the occurrence of an erasure could be used to trigger the return of a negative acknowledgement (NAK) to the originator of the message. The originator would react to the NAK by resending the erased frame.

8.2.3 State Table

Computer implementation of step #1 of Recipe 8.1 depends upon a state table that defines the particular code to be decoded. Table 8-12 is a state table for the code used in Examples 8.1 and 8.2. When designing the software implementation of such a table, we must consider the table from two different viewpoints—outside and inside. The stack decoding algorithm is simply a **user** of the table from the outside. The algorithm interrogates the table by providing values for present_state and input, and the table responds with the appropriate values of next_state and output. Any software construct capable of this behavior will satisfy the decoding algorithm's needs. Under an object-oriented philosophy, this is all that the decoder should know or care about the table. The exact internal details of how the table object generates the correct responses are intentionally hidden from external users of the table. The internal workings of the table can be implemented in any of the ways covered in Section 10. For purposes of the present section, we will assume that the interface to the state table is given by the following C prototypes:

```
/********************************************************/

int MealyEncoderNextState(int present_state, int input);

int MealyEncoderOutput(int present_state, int input);

/********************************************************/
```

Present state	Input	Next state	Output
0 (00)	0	0 (00)	00
0 (00)	1	2 (10)	11
1 (01)	0	0 (00)	11
1 (01)	1	2 (10)	00
2 (10)	0	1 (01)	10
2 (10)	1	3 (11)	01
3 (11)	0	1 (01)	01
3 (11)	1	3 (11)	10

8.3 Fano Decoding Algorithm

When the **stack** algorithm is working on a particular partial path that it "decides" is not likely to be part of the correct path (*i.e.,* when the metric of the partial path is inferior to metrics of one or more other partial paths in the stack) the algorithm jumps to the partial path that appears to be the one most likely to grow into the correct full path.

When the **Fano** algorithm "decides" that a particular partial path should be abandoned, a different approach is taken. Rather than jump to a completely different path, the algorithm "backs up" in the trellis by removing the most recently appended branch from the path. The algorithm then examines the alternative branches that could take the place of the one just removed.

To obtain its current working path, the **stack** algorithm selects the path with the best metric from among all partial paths that have been evaluated since the beginning of the codeword. Each one of these paths is stored, and after each iteration of the algorithm, it is a straightforward matter to select the one with the best metric.

The **Fano** algorithm stores only the one path from the initial node to the current working node, so it cannot use comparison with other paths to help decide what it should do with the current working path. Rather than looking for a better path, the Fano algorithm must decide if the current path is "good enough" to continue building upon. This decision is accomplished by comparing the new path metric to a threshold. If the new metric is better than the threshold, the algorithm moves along the corresponding new branch to reach the new working node. If the new metric is not better than the threshold, the algorithm examines alternate branching choices.

Several variants of the Fano algorithm have been described in the literature,[4] but I have found the easiest variant to understand and implement is the approach presented by Gallager. This approach is based upon the conditional application of the five rules listed in Table 8-13. The following paragraphs take a closer look at what the various items in the table actually mean.

■ **Table 8-13 Rules for the Fano decoding algorithm.**

	Conditions			Actions	
Rule	**Previous move**	M_n **compar.**	M_{n-1} **compar.**	**Final threshold**	**Move**
1	F or L	$M_n \geq T$	$M_{n-1} < T + \Delta$	Tighten	F
2	F or L	$M_n \geq T$	$M_{n-1} \geq T + \Delta$	No change	F
3	F or L	$M_n < T$	don't care	No change	L or B
4	B	don't care	$M_{n-1} < T$	Decrease by Δ	F
5	B	don't care	$M_{n-1} \geq T$	No change	L or B

The algorithm is fundamentally an approach for constructing (at the receiver) an estimate of the path through the encoding trellis that was traversed at the transmitter. This estimate is built by starting at the origin node and adding one branch at a time. The process of adding branches can be broken down into three different types of *moves*:

☐ A *forward move* occurs whenever a new branch is added to the partial tree under construction. If the "current" node before the move was at level n in the tree, the "current" node after the move will be at level $n + 1$.

☐ A *lateral move* occurs whenever the algorithm finds that the most recently added branch is unacceptable, and thereby determines that this branch must be removed and replaced with one of the alternative branches extending forward from the current node. If the "current" node before the move was at level n in the tree, the "current" node after the move will also be at level n.

---

4 Lin and D. J. Costello: *Error Control Coding: Fundamentals and Applications*, Prentice-Hall, Englewood Cliffs, NJ, 1983.

R. M. Fano, "A Heuristic Discussion of Probabilistic Decoding," *IEEE Trans. Inf. Theory*, IT-9, pp. 64-74, April 1963.

☐ A *backward move* occurs whenever the most recently added branch is found to be unacceptable and no untried alternative branches remain. The algorithm must then back up by one level and consider alternative forward branches. If the "current" node before the move was at level n in the tree, the "current" node after the move will be at level $n - 1$.

As indicated in the table, the direction of the **previous** move is one of the conditions used for selecting the rule to be applied. The other conditions for selection are based on comparison of path metrics to a threshold. Loosely speaking, a forward move is considered "good enough" to keep if the resulting path metric exceeds some threshold.[5] If the resulting path metric does not exceed the threshold, the algorithm attempts to find a better path. In the table, M_n denotes the path metric up through the "current" node.

Example 8.8.4 Assume that $\Delta = 4$. Rank (by decreasing metric) the forward branches from state 00 at level 0:

branch metric	final state	hypoth. input bit
–3.396	00	0
–3.396	10	1

1. Consider the best forward branch from state 00 at level 0.
   - The two forward branches out of state 00 at level 0 each have a branch metric of $M_B = -3.396$. The topmost branch, corresponding to a hypothesized encoder input of 0, is arbitrarily taken as the best branch. This branch goes to state 00 at level 1.
   - By convention, at the first step of the algorithm, the direction of the previous move is assumed to be **forward**.
   - $M_1 = -3.396$ and $T = 0$. Therefore, $M_1 < T$.
   - Rule 3 applies:
     ~Move **lateral** to state 10 at level 1.

2. Consider the second-best forward branch from state 00 level 0:
   - The previous move was **lateral**.
   - The second-best forward branch, corresponding to a hypothesized encoder input of 1, goes to state 10 at level 1 and has a branch metric of $M_B = -3.396$.
   - $M_1 = -3.396$ and $T = 0$. Therefore, $M_1 < T$.

---

5 Discussions of the Fano algorithm will assume that when metrics are compared, the largest metric is the "best" metric. As noted previously, there are a few metrics for which the smallest metric is the "best" metric. In such cases, all of the comparisons in Table 8-13 should be reversed in direction.

- Rule 3 applies:

  ~No possible lateral moves remain, so move **backward** to level -1.

3. Consider the forward branch **into** state 00 at level 0. This situation is treated as a special case because the encoder tree doesn't really have a level -1. By convention, $M_{-1} = -\infty$, so Rule 4 will always apply when the level is -1.
   - The previous move was **backward**.
   - $M_{-1} = -\infty$ and $T = 0$. Therefore, $M_{-1} < T$.
   - Rule 4 applies:

     ~Reduce threshold by Δ to $T = -4.0$.

     ~Move **forward** to state 00 at level 0.

     ~Rank (by decreasing metric) the forward branches from state 00 at level 1:

branch metric	final state	hypoth. input bit
-3.396	00	0
-3.396	10	1

4. Consider the best forward branch from state 00 at level 0.
   - The two forward branches out of state 00 at level 0 each have a branch metric of $M_B = -3.396$. The topmost branch, corresponding to a hypothesized encoder input of 0, is arbitrarily taken as the best branch. This branch goes to state 00 at level 1.
   - The previous move was **forward**.
   - $M_1 = -3.396$ and $T = -4$. Therefore, $M_1 \geq T$.
   - $M_0 = 0$ and $T + \Delta = 0$. Therefore, $M_0 \geq T + 0$.
   - Rule 2 applies:

     ~Move forward to state 00 at level 1.

     ~Rank (by decreasing metric) the forward branches from state 00 at level 1:

branch metric	final state	hypoth. input bit
-3.396	00	0
-3.396	10	1

5. Consider the best forward branch from state 00 at level 1.
   - The two forward branches out of state 00 at level 1 each have a branch metric of $M_B = -3.396$. The topmost branch, corresponding to a hypothesized encoder input of 0, is arbitrarily taken as the best branch. This branch goes to state 00 at level 2.

- The previous move was **forward**.
- $M_2 = -6.792$ and $T = -4$. Therefore, $M_2 < T$.
- Rule 3 applies:

    ~Move **lateral** to state 10 at level 2.

6. Consider the second-best forward branch from state 00 level 1:
    - The previous move was **lateral**.
    - The second-best forward branch, corresponding to a hypothesized encoder input of 1, goes to state 10 at level 2 and has a branch metric of $M_B = -3.396$.
    - $M_2 = -6.792$ and $T = -4.0$. Therefore, $M_2 < T$.
    - Rule 3 applies:

        ~No possible lateral moves remain, so move **backward** to state 00 at level 0.

        ~Rank (by decreasing metric) the forward branches from state 00 at level 0:

branch metric	final state	hypoth. input bit
-3.396	00	0
-3.396	10	1

7. Consider the best forward branch from state 00 at level 0.
    - The two forward branches out of state 00 at level 1 each have a branch metric of $M_B = -3.396$. The topmost branch, corresponding to a hypothesized encoder input of 0, is arbitrarily taken as the best branch. This branch goes to state 00 at level 1.
    - The previous move was **backward**.
    - $M_0 = 0$ and $T = -4$. Therefore, $M_0 \geq T$.
    - Rule 5 applies:

        ~Move **lateral** to state 10 at level 1.

8. Consider the second-best forward branch from state 00 at level 0:
    - The previous move was **lateral**.
    - The second-best forward branch, corresponding to a hypothesized encoder input of 1, goes to state 10 at level 1 and has a branch metric of $M_B = -3.396$.
    - $M_1 = -3.396$ and $T = -4.0$. Therefore, $M_1 \geq T$.
    - $M_0 = 0$ and $T + \Delta = 0$. Therefore, $M_0 \geq T + \Delta$.
    - Rule 2 applies:

        ~Move **forward** to state 10 at level 1.

        ~Rank (by decreasing metric) the forward branches from state 10 at level 1:

184

branch metric	final state	hypoth. input bit
0.852	01	0
−7.644	11	1

9. Consider the best forward branch from state 10 at level 1:
   - The previous move was **forward**.
   - The best forward branch, corresponding to a hypothesized encoder input of 0, goes to state 01 at level 2 and has a branch metric of $M_B = 0.852$.
   - The hypothesized partial input sequence at this point is 1, 0.
   - $M_2 = -2.544$ and $T = -4.0$. Therefore, $M_2 \geq T$.
   - $M_1 = -3.396$ and $T + \Delta = 0$. Therefore, $M_1 < T + \Delta$.
   - Rule 1 applies:

     ~Attempt unsuccessfully to tighten threshold, leaving $T = -4.0$.

     ~Move **forward** to state 01 at level 2.

     ~Rank (by decreasing metric) the forward branches from state 01 at level 2:

branch metric	final state	hypoth. input bit
−7.644	00	0
0.852	10	1

10. Consider the best forward branch from state 01 at level 2:
    - The previous move was **forward**.
    - The best forward branch, corresponding to a hypothesized encoder input of 1, goes to state 10 at level 3 and has a branch metric of $M_B = 0.852$.
    - The hypothesized partial input sequence at this point is 1, 0, 1.
    - $M_3 = -1.692$ and $T = -4.0$. Therefore, $M_3 \geq T$.
    - $M_2 = -2.544$ and $T + \Delta = 0$. Therefore, $M_2 < T + \Delta$.
    - Rule 1 applies:

      ~Attempt unsuccessfully to tighten threshold, leaving $T = -4.0$.

      ~Move **forward** to state 10 at level 3.

      ~Rank (by decreasing metric) the forward branches from state 10 at level 3:

branch metric	final state	hypoth. input bit
0.852	01	0
−7.644	11	1

11. Consider the best forward branch from state 10 at level 3:
    - The previous move was **forward**.
    - The best forward branch, corresponding to a hypothesized encoder input of 0, goes to state 01 at level 4 and has a branch metric of $M_B = 0.852$.
    - The hypothesized partial input sequence at this point is 1, 0, 1, 0.
    - $M_4 = -0.84$ and $T = -4.0$. Therefore, $M_4 \geq T$.
    - $M_3 = -1.692$ and $T + \Delta = 0$. Therefore, $M_3 < T + \Delta$.
    - Rule 1 applies:
      ~Attempt unsuccessfully to tighten threshold, leaving $T = -4.0$.
      ~Move **forward** to state 01 at level 4.
      ~Rank (by decreasing metric) the forward branches from state 01 at level 4.

branch metric	final state	hypoth. input bit
0.852	00	0
−7.644	10	1

12. Consider the best forward branch from state 01 at level 4:
    - The previous move was **forward**.
    - The best forward branch, corresponding to a hypothesized encoder input of 0, goes to state 00 at level 5 and has a branch metric of $M_B = 0.852$.
    - The hypothesized partial input sequence at this point is 1, 0, 1, 0, 0.
    - $M_5 = 0.012$ and $T = -4.0$. Therefore, $M_4 \geq T$.
    - $M_4 = -0.84$ and $T + \Delta = 0$. Therefore, $M_3 < T + \Delta$.
    - Rule 1 applies:
      ~Tighten threshold to $T = 0$.
      ~Move **forward** to state 00 at level 5.
      ~Rank (by decreasing metric) the forward branches from state 00 at level 5:

branch metric	final state	hypoth. input bit
−3.396	00	0
−3.396	10	1

13. Consider the best forward branch from state 00 at level 5:
    - The previous move was **forward**.
    - The two forward branches out of state 00 at level 5 each have a branch metric of $M_B = -3.396$. The topmost branch, corresponding to a hypothesized encoder input of 0, is arbitrarily taken as the best branch. This branch goes to state 00 at level 6.
    - The hypothesized partial input sequence at this point is 1, 0, 1, 0, 0, 0.
    - $M_6 = -3.384$ and $T = 0$. Therefore, $M_6 < T$.
    - Rule 3 applies:
      ~Move **lateral** to state 10 at level 6.

14. Consider the second-best forward branch from state 00 at level 5:
    - The previous move was **lateral**.
    - The second-best forward branch, corresponding to a hypothesized encoder input of 1, goes to state 10 at level 6 and has a branch metric of $M_B = -3.396$.
    - $M_6 = -3.384$ and $T = 0$. Therefore, $M_6 < T$.
    - Rule 3 applies:
      ~No possible lateral moves remain, so move **backward** to state 00 at level 4.
      ~Rank (by decreasing metric) the forward branches from state 00 at level 4:

branch metric	final state	hypoth. input bit
0.852	00	0
-7.644	10	1

15. Consider the best forward branch from state 00 at level 4:
    - The previous move was **backward**.
    - The best forward branch, corresponding to a hypothesized encoder input of 0, goes to state 00 at level 5 and has a branch metric of $M_B = 0.852$.
    - The hypothesized partial input sequence at this point is 1, 0, 1, 0, 0.
    - Rule 4 applies:
      ~Reduce threshold by Δ to $T = -4.0$.
      ~Move **forward** to state 00 at level 5.

~Rank (by decreasing metric) the forward branches from state 00 at level 5:

branch metric	final state	hypoth. input bit
−3.396	00	0
−3.396	10	1

16. Consider the best forward branch from state 00 at level 5:
    - The previous move was **forward**.
    - The two forward branches out of state 00 at level 5 each have a branch metric of $M_B = -3.396$. The topmost branch, corresponding to a hypothesized encoder input of 0, is arbitrarily taken as the best branch. This branch goes to state 00 at level 6.
    - The hypothesized partial input sequence at this point is 1, 0, 1, 0, 0, 0.
    - $M_6 = -3.384$ and $T = -4.0$. Therefore, $M_6 \geq T$.
    - $M_5 = 0.012$ and $T + \Delta = 0$. Therefore, $M_5 \geq T + \Delta$.
    - Rule 2 applies:

    ~Move **forward** to state 00 at level 6.

    ~Rank (by decreasing metric) the forward branches from state 00 at level 6:

branch metric	final state	hypoth. input bit
−7.644	00	0
0.852	10	1

17. Consider the best forward branch from state 00 at level 6:
    - The previous move was **forward**.
    - The best forward branch, corresponding to a hypothesized encoder input of 1, goes to state 10 at level 7 and has a branch metric of $M_B = 0.852$.
    - The hypothesized partial input sequence at this point is 1, 0, 1, 0, 0, 0, 1.
    - $M_7 = -2.532$ and $T = -4.0$. Therefore, $M_7 \geq T$.
    - $M_6 = -3.384$ and $T + \Delta = 0$. Therefore, $M_6 < T + \Delta$.
    - Rule 1 applies:

    ~Attempt unsuccessfully to tighten threshold, leaving $T = -4.0$.

    ~Move **forward** to state 10 at level 7.

    ~Rank (by decreasing metric) the forward branches from state 10 at level 7:

branch metric	final state	hypoth. input bit
0.852	01	0
−7.644	11	1

18. Consider the best forward branch from state 10 at level 7:
    - The previous move was **forward**.
    - The best forward branch, corresponding to a hypothesized encoder input of 0, goes to state 01 at level 8 and has a branch metric of $M_B = 0.852$.
    - The hypothesized partial input sequence at this point is 1, 0, 1, 0, 0, 0, 1, 0.
    - $M_8 = -1.68$ and $T = -4.0$. Therefore, $M_8 \geq T$.
    - $M_7 = -2.532$ and $T + \Delta = 0$. Therefore, $M_7 < T + \Delta$.
    - Rule 1 applies:
      - ~Attempt unsuccessfully to tighten threshold, leaving $T = -4.0$.
      - ~Move forward to state 01 at level 8.
      - ~Rank (by decreasing metric) the forward branches from state 01 at level 8.

branch metric	final state	hypoth. input bit
0.852	00	0
−7.644	10	1

19. Consider the best forward branch from state 01 at level 8:
    - The previous move was **forward**.
    - The best forward branch, corresponding to a hypothesized encoder input of 0, goes to state 00 at level 9 and has a branch metric of $M_B = 0.852$.
    - The hypothesized partial input sequence at this point is 1, 0, 1, 0, 0, 0, 1, 0, 0.
    - $M_9 = -0.828$ and $T = -4.0$. Therefore, $M_9 \geq T$.
    - $M_8 = -1.68$ and $T + \Delta = 0$. Therefore, $M_8 < T + \Delta$.
    - Rule 1 applies:
      - ~Attempt unsuccessfully to tighten threshold, leaving $T = -4.0$.
      - ~Move **forward** to state 00 at level 9.
      - ~Decoding is complete, and the hypothesized input sequence is correct.

Minimal Polynomials of Elements in GF(2^m) for Chapter 2

A

■ **Table 2A-1 Minimal polynomials $\phi_i(x)$ of elements α_i in GF(2^2).**

i	powers of nonzero terms in $\phi_i(x)$
1	$(2, 1, 0)$

■ **Table 2A-2 Minimal polynomials $\phi_i(x)$ of elements α_i in GF(2^3)**

i	powers of nonzero terms in $\phi_i(x)$
1	$(3, 1, 0)$
3	$(3, 2, 0)$

■ **Table 2A-3 Minimal polynomials $\phi_i(x)$ of elements α_i in GF(2^4).**

i	powers of nonzero terms in $\phi_i(x)$
1	$(4, 1, 0)$
3	$(4, 3, 2, 1, 0)$
5	$(2, 1, 0)$
7	$(4, 3, 0)$

■ Table 2A-4
Minimal polynomials
$\phi_i(x)$ of elements
$\alpha_i(x)$ in GF(2^5).

i	powers of nonzero terms in $\phi_i(x)$
1	(5, 2, 0)
3	(5, 4, 3, 2, 0)
5	(5, 4, 2, 1, 0)
7	(5, 3, 2, 1, 0)
11	(5, 4, 3, 1, 0)
15	(5, 3, 0)

■ Table 2A-5 Minimal polynomials
$\phi_i(x)$ of elements $\alpha_i(x)$ in GF(2^6).

i	powers of nonzero terms in $\phi_i(x)$	i	powers of nonzero terms in $\phi_i(x)$
1	(6, 1, 0)	13	(6, 4, 3, 1, 0)
3	(6, 4, 2, 1, 0)	15	(6, 5, 4, 2, 0)
5	(6, 5, 2, 1, 0)	21	(2, 1, 0)
7	(6, 3, 0)	23	(6, 5, 4, 1, 0)
9	(3, 2, 0)	27	(3, 1, 0)
11	(6, 5, 3, 2, 0)	31	(6, 5, 0)

■ Table 2A-6 Minimal polynomials
$\phi_i(x)$ of elements $\alpha_i(x)$ in GF(2^7).

i	powers of nonzero terms in $\phi_i(x)$	i	powers of nonzero terms in $\phi_i(x)$
1	(7, 3, 0)	21	(7, 6, 5, 2, 0)
3	(7, 3, 2, 1, 0)	23	(7, 6, 0)
5	(7, 4, 3, 2, 0)	27	(7, 6, 4, 1, 0)
7	(7, 6, 5, 4, 2, 1, 0)	29	(7, 5, 3, 1, 0)
9	(7, 5, 4, 3, 2, 1, 0)	31	(7, 6, 5, 4, 0)
11	(7, 6, 4, 2, 0)	43	(7, 5, 2, 1, 0)
13	(7, 1, 0)	47	(7, 5, 4, 3, 0)
15	(7, 6, 5, 3, 2, 1, 0)	55	(7, 6, 5, 4, 3, 2, 0)
19	(7, 6, 2, 1, 0)	63	(7, 4, 0)

■ Table 2A-7 Minimal polynomials $\phi_i(x)$ of elements $\alpha_i(x)$ in GF(2^8).

i	powers of nonzero terms in $\phi_i(x)$	i	powers of nonzero terms in $\phi_i(x)$
1	(8, 4, 3, 2, 0)	39	(8, 7, 6, 5, 4, 3, 0)
3	(8, 6, 5, 4, 2, 1, 0)	43	(8, 7, 6, 1, 0)
5	(8, 7, 6, 5, 4, 1, 0)	45	(8, 5, 4, 3, 0)
7	(8, 6, 5, 3, 0)	47	(8, 7, 5, 3, 0)
9	(8, 7, 5, 4, 3, 2, 0)	51	(4, 3, 2, 1, 0)
11	(8, 7, 6, 5, 2, 1, 0)	53	(8, 7, 2, 1, 0)
13	(8, 5, 3, 1, 0)	55	(8, 7, 5, 4, 0)
15	(8, 7, 6, 4, 2, 1, 0)	59	(8, 6, 3, 2, 0)
17	(4, 1, 0)	61	(8, 7, 6, 3, 2, 1, 0)
19	(8, 6, 5, 2, 0)	63	(8, 7, 6, 4, 3, 2, 0)
21	(8, 7, 3, 1, 0)	85	(2, 1, 0)
23	(8, 6, 5, 1, 0)	87	(8, 7, 5, 1, 0)
25	(8, 4, 3, 1, 0)	91	(8, 7, 6, 5, 4, 2, 0)
27	(8, 5, 4, 3, 2, 1, 0)	95	(8, 7, 4, 3, 2, 1, 0)
29	(8, 7, 3, 2, 0)	111	(8, 6, 5, 4, 3, 1, 0)
31	(8, 5, 3, 2, 0)	119	(4, 3, 0)
37	(8, 6, 4, 3, 2, 1, 0)	127	(8, 6, 5, 4, 0)

■ Table 2A-8 Minimal polynomials $\phi_i(x)$ of elements $\alpha_i(x)$ in GF(2^9).

i	powers of nonzero terms in $\phi_i(x)$	i	powers of nonzero terms in $\phi_i(x)$
1	(9, 4, 0)	31	(9, 4, 3, 1, 0)
3	(9, 6, 4, 3, 0)	35	(9, 8, 0)
5	(9, 8, 5, 4, 0)	37	(9, 6, 5, 3, 2, 1, 0)
7	(9, 7, 4, 3, 0)	39	(9, 8, 7, 6, 3, 2, 0)
9	(9, 8, 4, 1, 0)	41	(9, 8, 6, 5, 4, 1, 0)
11	(9, 5, 3, 2, 0)	43	(9, 8, 7, 6, 3, 1, 0)
13	(9, 6, 5, 4, 2, 1, 0)	45	(9, 6, 5, 4, 3, 2, 0)
15	(9, 8, 6, 5, 0)	47	(9, 8, 6, 4, 3, 1, 0)
17	(9, 7, 6, 4, 3, 1, 0)	51	(9, 8, 7, 6, 4, 2, 0)
19	(9, 8, 7, 2, 0)	53	(9, 7, 4, 2, 0)
21	(9, 4, 2, 1, 0)	55	(9, 7, 5, 4, 3, 2, 0)
23	(9, 8, 7, 6, 5, 3, 0)	57	(9, 7, 6, 5, 4, 2, 0)
25	(9, 8, 7, 6, 5, 1, 0)	59	(9, 7, 6, 3, 2, 1, 0)
27	(9, 8, 7, 3, 2, 1, 0)	61	(9, 6, 4, 3, 2, 1, 0)
29	(9, 8, 6, 5, 3, 1, 0)	63	(9, 6, 5, 2, 0)

i	powers of nonzero terms in $\phi_i(x)$	i	powers of nonzero terms in $\phi_i(x)$
73	$(3, 1, 0)$	117	$(9, 8, 6, 3, 2, 1, 0)$
75	$(9, 8, 7, 6, 5, 4, 3, 1, 0)$	119	$(9, 1, 0)$
77	$(9, 8, 6, 3, 0)$	123	$(9, 7, 2, 1, 0)$
79	$(9, 8, 7, 6, 2, 1, 0)$	125	$(9, 7, 6, 4, 0)$
83	$(9, 8, 4, 2, 0)$	127	$(9, 6, 5, 3, 0)$
85	$(9, 7, 6, 4, 2, 1, 0)$	171	$(9, 8, 7, 5, 4, 2, 0)$
87	$(9, 7, 5, 2, 0)$	175	$(9, 8, 7, 5, 0)$
91	$(9, 6, 3, 1, 0)$	183	$(9, 8, 5, 3, 1, 0)$
93	$(9, 7, 6, 5, 4, 3, 0)$	187	$(9, 8, 7, 6, 4, 3, 0)$
95	$(9, 8, 7, 5, 4, 3, 0)$	191	$(9, 5, 4, 1, 0)$
103	$(9, 7, 5, 3, 2, 1, 0)$	219	$(3, 2, 0)$
107	$(9, 7, 5, 1, 0)$	223	$(9, 8, 5, 1, 0)$
109	$(9, 8, 6, 5, 4, 3, 2, 1, 0)$	239	$(9, 8, 6, 5, 3, 2, 0)$
111	$(9, 8, 4, 3, 2, 1, 0)$	255	$(9, 5, 0)$

■ Table 2A-9 Minimal polynomials $\phi_i(x)$ of elements $\alpha_i(x)$ in GF(2^{10}).

i	powers of nonzero terms in $\phi_i(x)$	i	powers of nonzero terms in $\phi_i(x)$
1	$(10, 3, 0)$	41	$(10, 8, 7, 6, 5, 2, 0)$
3	$(10, 3, 2, 1, 0)$	43	$(10, 8, 4, 3, 0)$
5	$(10, 8, 3, 2, 0,)$	45	$(10, 9, 5, 4, 0)$
7	$(10, 9, 8, 7, 6, 5, 4, 3, 0)$	47	$(10, 9, 6, 5, 4, 3, 2, 1, 0)$
9	$(10, 7, 5, 3, 2, 1, 0)$	49	$(10, 9, 8, 6, 4, 2, 0)$
11	$(10, 5, 4, 2, 0)$	51	$(10, 8, 6, 5, 2, 1, 0)$
13	$(10, 6, 5, 3, 2, 1, 0)$	53	$(10, 8, 7, 3, 2, 1, 0)$
15	$(10, 8, 7, 5, 3, 1, 0)$	55	$(10, 9, 8, 5, 3, 1, 0)$
17	$(10, 9, 8, 6, 3, 2, 0)$	57	$(10, 9, 6, 4, 0)$
19	$(10, 8, 7, 6, 5, 4, 3, 1, 0)$	59	$(10, 9, 8, 5, 4, 3, 0)$
21	$(10, 9, 8, 7, 6, 5, 3, 1, 0)$	61	$(10, 9, 8, 7, 6, 5, 4, 1, 0)$
23	$(10, 4, 3, 1, 0)$	63	$(10, 9, 7, 5, 3, 2, 0)$
25	$(10, 8, 5, 1, 0)$	69	$(10, 8, 7, 6, 0)$
27	$(10, 9, 8, 6, 5, 4, 3, 1, 0)$	71	$(10, 9, 7, 6, 4, 1, 0)$
29	$(10, 8, 5, 4, 0)$	73	$(10, 9, 8, 6, 2, 1, 0)$
31	$(10, 9, 5, 1, 0)$	75	$(10, 8, 4, 3, 2, 1, 0)$
33	$(5, 4, 3, 2, 0)$	77	$(10, 8, 3, 1, 0)$
35	$(10, 9, 4, 1, 0)$	79	$(10, 7, 6, 5, 2, 1, 0)$
37	$(10, 9, 8, 6, 5, 1, 0)$	83	$(10, 9, 8, 7, 4, 1, 0)$
39	$(10, 6, 2, 1, 0)$	85	$(10, 8, 7, 6, 2, 1, 0)$

i	powers of nonzero terms in $\phi_i(x)$	i	powers of nonzero terms in $\phi_i(x)$
87	$(10, 7, 6, 3, 0)$	187	$(10, 9, 7, 2, 0)$
89	$(10, 7, 6, 4, 2, 1, 0)$	189	$(10, 6, 5, 1, 0)$
91	$(10, 9, 7, 5, 4, 2, 0)$	191	$(10, 9, 8, 7, 5, 4, 0)$
93	$(10, 9, 8, 7, 6, 5, 4, 3, 2, 1, 0)$	205	$(10, 7, 3, 1, 0)$
95	$(10, 6, 5, 2, 0)$	207	$(10, 9, 8, 5, 4, 2, 0)$
99	$(5, 4, 2, 1, 0)$	213	$(10, 8, 7, 4, 3, 1, 0)$
101	$(10, 5, 3, 2, 0)$	215	$(10, 8, 7, 5, 0)$
103	$(10, 9, 8, 6, 5, 4, 3, 2, 0)$	219	$(10, 8, 7, 5, 4, 3, 0)$
105	$(10, 9, 8, 7, 2, 1, 0)$	221	$(10, 9, 8, 6, 4, 3, 0)$
107	$(10, 9, 6, 5, 4, 3, 0)$	223	$(10, 9, 5, 2, 0)$
109	$(10, 5, 2, 1, 0)$	231	$(5, 4, 3, 1, 0)$
111	$(10, 6, 4, 1, 0)$	235	$(10, 9, 6, 3, 2, 1, 0)$
115	$(10, 8, 7, 6, 5, 4, 2, 1, 0)$	237	$(10, 9, 8, 7, 6, 2, 0)$
117	$(10, 7, 4, 3, 0)$	239	$(10, 8, 6, 4, 2, 1, 0)$
119	$(10, 9, 6, 4, 3, 1, 0)$	245	$(10, 7, 6, 2, 0)$
121	$(10, 9, 7, 5, 2, 1, 0)$	247	$(10, 9, 6, 1, 0)$
123	$(10, 9, 8, 4, 0)$	251	$(10, 9, 7, 6, 5, 4, 3, 2, 0)$
125	$(10, 9, 7, 6, 0)$	253	$(10, 8, 6, 5, 0)$
127	$(10, 7, 6, 5, 4, 3, 2, 1, 0)$	255	$(10, 9, 8, 7, 0)$
147	$(10, 7, 6, 5, 3, 2, 0)$	341	$(2, 1, 0)$
149	$(10, 9, 4, 2, 0)$	343	$(10, 9, 8, 4, 3, 2, 0)$
151	$(10, 9, 8, 5, 0)$	347	$(10, 8, 6, 1, 0)$
155	$(10, 7, 5, 3, 0)$	351	$(10, 9, 7, 5, 4, 3, 2, 1, 0)$
157	$(10, 8, 6, 5, 3, 1, 0)$	363	$(5, 2, 0)$
159	$(10, 9, 7, 6, 5, 4, 2, 1, 0)$	367	$(10, 8, 5, 4, 3, 2, 0)$
165	$(5, 3, 0)$	375	$(10, 4, 3, 2, 0)$
167	$(10, 7, 6, 5, 4, 1, 0)$	379	$(10, 9, 5, 4, 2, 1, 0)$
171	$(10, 9, 7, 6, 3, 2, 0)$	383	$(10, 8, 7, 2, 0)$
173	$(10, 9, 7, 6, 4, 3, 2, 1, 0)$	439	$(10, 9, 8, 4, 2, 1, 0)$
175	$(10, 8, 7, 3, 2, 0)$	447	$(10, 9, 8, 7, 5, 3, 0)$
179	$(10, 9, 7, 3, 0)$	479	$(10, 8, 7, 4, 2, 1, 0)$
181	$(10, 9, 8, 7, 6, 4, 3, 1, 0)$	495	$(5, 3, 2, 1, 0)$
183	$(10, 9, 8, 3, 2, 1, 0)$	511	$(10, 7, 0)$

Stack Tables
for Example 8.1

B

■ **Table 8B.1** Stack contents after first processing cycle.

Inputs	Final encoder state	Path metric
†1	2	0
0	0	2

■ **Table 8B.2** Stack contents after second processing cycle.

Inputs	Final encoder state	Path metric
†*10	1	1
*11	3	1
0	0	2

■ **Table 8B.3** Stack contents after third processing cycle.

Inputs	Final encoder state	Path metric
†*101	2	1
11	3	1
0	0	2
*100	0	3

Inputs	Final encoder state	Path metric
11	3	1
†*1010	1	2
*1011	3	2
0	0	2
100	0	3

■ Table 8B.5 Stack contents
after processing cycle 5.

Inputs	Final encoder state	Path metric
†1010	1	2
1011	3	2
*110	1	2
*111	3	2
0	0	2
100	0	3

■ Table 8B.6 Stack contents
after processing cycle 6.

Inputs	Final encoder state	Path metric
†*10100	0	2
1011	3	2
110	1	2
111	3	2
0	0	2
100	0	3
*10101	2	4

■ Table 8B.7 Stack contents after processing cycle 7.

Inputs	Final encoder state	Path metric
†*101000	0	2
1011	3	2
110	1	2
111	3	2
0	0	2
100	0	3
*101001	2	4
10101	2	4

■ Table 8B.8 Stack contents after processing cycle 8.

Inputs	Final encoder state	Path metric
1011	3	2
110	1	2
111	3	2
0	0	2
*1010000	0	3
†*1010001	2	3
100	0	3
101001	2	4
10101	2	4

■ Table 8B.9 Stack
contents after processing cycle 9.

Inputs	Final encoder state	Path metric
110	1	2
111	3	2
0	0	2
1010000	0	3
†1010001	2	3
*10110	1	3
*10111	3	3
100	0	3
101001	2	4
10101	2	4

■ Table 8B.10 Stack
contents after processing cycle 10.

Inputs	Final encoder state	Path metric
*1101	2	2
111	3	2
0	0	2
1010000	0	3
†1010001	2	3
10110	1	3
10111	3	3
100	0	3
101001	2	4
10101	2	4
*1100	0	4

■ **Table 8B.11** Stack
contents after processing cycle 11.

Inputs	Final encoder state	Path metric
111	3	2
0	0	2
1010000	0	3
†1010001	2	3
*11010	1	3
*11011	3	3
10110	1	3
10111	3	3
100	0	3
101001	2	4
10101	2	4
1100	0	4

■ **Table 8B.12** Stack
contents after processing cycle 12.

Inputs	Final encoder state	Path metric
0	0	2
1010000	0	3
†1010001	2	3
11010	1	3
11011	3	3
10110	1	3
10111	3	3
*1110	1	3
*1111	3	3
100	0	3
101001	2	4
10101	2	4
1100	0	4

■ Table 8B.13 Stack
contents after processing cycle 13.

Inputs	Final encoder state	Path metric
*00	0	2
1010000	0	3
†1010001	2	3
11010	1	3
11011	3	3
10110	1	3
10111	3	3
1110	1	3
1111	3	3
100	0	3
101001	2	4
10101	2	4
1100	0	4
*01	2	4

■ Table 8B.14 Stack
contents after processing cycle 14.

Inputs	Final encoder state	Path metric
*000	0	2
1010000	0	3
†1010001	2	3
11010	1	3
11011	3	3
10110	1	3
10111	3	3
1110	1	3
1111	3	3
100	0	3
101001	2	4
10101	2	4
1100	0	4
*001	2	4
01	2	4

■ Table 8B.15 Stack contents after processing cycle 15.

Inputs	Final encoder state	Path metric
*0000	0	2
1010000	0	3
†1010001	2	3
11010	1	3
11011	3	3
10110	1	3
10111	3	3
1110	1	3
1111	3	3
100	0	3
101001	2	4
10101	2	4
*0001	2	4
1100	0	4
001	2	4
01	2	4

■ Table 8B.16 Stack contents after processing cycle 16.

Inputs	Final encoder state	Path metric
*00001	2	2
1010000	0	3
†1010001	2	3
11010	1	3
11011	3	3
10110	1	3
10111	3	3
1110	1	3
1111	3	3
100	0	3
101001	2	4
*00000	0	4
10101	2	4
0001	2	4
1100	0	4
001	2	4
01	2	4

■ Table 8B.17 Stack
contents after processing cycle 17.

Inputs	Final encoder state	Path metric
1010000	0	3
†1010001	2	3
*000010	1	3
*000011	3	3
11010	1	3
11011	3	3
10110	1	3
10111	3	3
1110	1	3
1111	3	3
100	0	3
101001	2	4
00000	0	4
10101	2	4
0001	2	4
1100	0	4
001	2	4
01	2	4

■ Table 8B.18 Stack
contents after processing cycle 18.

Inputs	Final encoder state	Path metric
†1010001	2	3
000010	1	3
000011	3	3
11010	1	3
11011	3	3
10110	1	3
10111	3	3
1110	1	3
1111	3	3
100	0	3
*10100000	0	4
101001	2	4
00000	0	4
10101	2	4
0001	2	4

Stack Tables for Example 8.1

■ Table 8B.18 Continued.

Inputs	Final encoder state	Path metric
1100	0	4
001	2	4
01	2	4

■ Table 8B.19 Stack contents after processing cycle 19.

Inputs	Final encoder state	Path metric
†*10100010	1	3
000010	1	3
000011	3	3
11010	1	3
11011	3	3
10110	1	3
10111	3	3
1110	1	3
1111	3	3
100	0	3
10100000	0	4
101001	2	4
00000	0	4
10101	2	4
0001	2	4
1100	0	4
001	2	4
01	2	4

■ Table 8B.20 Stack
contents after processing cycle 20.

Inputs	Final encoder state	Path metric
†*101000100	0	3
000010	1	3
000011	3	3
11010	1	3
11011	3	3
10110	1	3
10111	3	3
1110	1	3
1111	3	3
100	0	3
10100000	0	4
101001	2	4
00000	0	4
10101	2	4
0001	2	4
1100	0	4
001	2	4
01	2	4

Stack Tables
for Example 8.2

C

■ **Table 8C.1** Stack
contents for Example 8.2
after first processing cycle.

Inputs	Final encoder state	Path metric
†1	2	0.852
0	0	−7.644

■ **Table 8C.2** Stack
contents for Example 8.2
after second processing cycle.

Inputs	Final encoder state	Path metric
†*10	1	−2.544
*11	3	−2.544
0	0	−7.644

■ **Table 8C.3** Stack
contents for Example
8.2 after processing cycle 3.

Inputs	Final encoder state	Path metric
†*101	2	−1.692
11	3	−2.544
0	0	−7.644
*100	0	−10.188

■ Table 8C.4 Stack
contents for Example
8.2 after processing cycle 4.

Inputs	Final encoder state	Path metric
11	3	−2.544
†*1010	1	−5.088
*1011	3	−5.088
0	0	−7.644
100	0	−10.188

■ Table 8C.5 Stack
contents for Example
8.2 after processing cycle 5.

Inputs	Final encoder state	Path metric
†1010	1	−5.088
1011	3	−5.088
*110	1	−5.94
*111	3	−5.94
0	0	−7.644
100	0	−10.188

■ Table 8C.6 Stack
contents for Example
8.2 after processing cycle 6.

Inputs	Final encoder state	Path metric
†*10100	0	−4.236
1011	3	−5.088
110	1	−5.94
111	3	−5.94
0	0	−7.644
100	0	−10.188
*10101	2	−12.732

■ Table 8C.7 Stack contents for Example 8.2 after processing cycle 7.

Inputs	Final encoder state	Path metric
†*101000	0	−3.384
1011	3	−5.088
110	1	−5.94
111	3	−5.94
0	0	−7.644
100	0	−10.188
*101001	2	−11.88
10101	2	−12.732

■ Table 8C.8 Stack contents for Example 8.2 after processing cycle 8.

Inputs	Final encoder state	Path metric
1011	3	−5.088
110	1	−5.94
111	3	−5.94
*1010000	0	−6.78
†*1010001	2	−6.78
0	0	−7.644
100	0	−10.188
101001	2	−11.88
10101	2	−12.732

Table 8C.9 Stack contents for Example 8.2 after processing cycle 9.

Inputs	Final encoder state	Path metric
110	1	−5.94
111	3	−5.94
1010000	0	−6.78
†1010001	2	−6.78
0	0	−7.644
*10110	1	−8.484
*10111	3	−8.484
100	0	−10.188
101001	2	−11.88
10101	2	−12.732

Table 8C.10 Stack contents for Example 8.2 after processing cycle 10.

Inputs	Final encoder state	Path metric
*1101	2	−5.088
111	3	−5.94
1010000	0	−6.78
†1010001	2	−6.78
0	0	−7.644
10110	1	−8.484
10111	3	−8.484
100	0	−10.188
101001	2	−11.88
10101	2	−12.732
*1100	0	−13.584

■ Table 8C.11 Stack
contents for Example
8.2 after processing cycle 11.

Inputs	Final encoder state	Path metric
111	3	−5.94
1010000	0	−6.78
†1010001	2	−6.78
0	0	−7.644
*11010	1	−8.484
*11011	3	−8.484
10110	1	−8.484
10111	3	−8.484
100	0	−10.188
101001	2	−11.88
10101	2	−12.732
1100	0	−13.584

■ Table 8C.12 Stack
contents for Example
8.2 after processing cycle 12.

Inputs	Final encoder state	Path metric
1010000	0	−6.78
†1010001	2	−6.78
0	0	−7.644
11010	1	−8.484
11011	3	−8.484
10110	1	−8.484
10111	3	−8.484
*1110	1	−9.336
*1111	3	−9.336
100	0	−10.188
101001	2	−11.88
10101	2	−12.732
1100	0	−13.584

■ Table 8C.13 Stack
contents for Example
8.2 after processing cycle 13.

Inputs	Final encoder state	Path metric
†1010001	2	−6.78
0	0	−7.644
11010	1	−8.484
11011	3	−8.484
10110	1	−8.484
10111	3	−8.484
1110	1	−9.336
1111	3	−9.336
*10100000	0	−10.176
100	0	−10.188
101001	2	−11.88
10101	2	−12.732
1100	0	−13.584

■ Table 8C.14 Stack
contents for Example
8.2 after processing cycle 14.

Inputs	Final encoder state	Path metric
†*10100010	1	−5.928
0	0	−7.644
11010	1	−8.484
11011	3	−8.484
10110	1	−8.484
10111	3	−8.484
1110	1	−9.336
1111	3	−9.336
10100000	0	−10.176
100	0	−10.188
101001	2	−11.88
10101	2	−12.732
1100	0	−13.584

**■ Table 8C.15 Stack
contents for Example
8.2 after processing cycle 15.**

Inputs	Final encoder state	Path metric
†*101000100	0	−5.076
0	0	−7.644
11010	1	−8.484
11011	3	−8.484
10110	1	−8.484
10111	3	−8.484
1110	1	−9.336
1111	3	−9.336
10100000	0	−10.176
100	0	−10.188
101001	2	−11.88
10101	2	−12.732
1100	0	−13.584

Stack Tables
for Example 8.3

D

■ Table 8D.1
Stack contents after
the first processing
cycle for the 5-error case.

Inputs	Final encoder state	Path metric
*0	0	−3.396
*1	2	−3.396

■ Table 8D.2 Stack
contents for the 5-error
case after processing cycle 2.

Inputs	Final encoder state	Path metric
1	2	−3.396
*00	0	−6.792
*01	2	−6.792

■ Table 8D.3 Stack
contents for the 5-error
case after processing cycle 3.

Inputs	Final encoder state	Path metric
*10	1	−2.544
00	0	−6.792
01	2	−6.792
*11	3	−11.04

■ Table 8D.4 Stack
contents for the 5-error
case after processing cycle 4.

Inputs	Final encoder state	Path metric
*100	0	−1.692
00	0	−6.792
01	2	−6.792
*101	2	−10.188
11	3	−11.04

■ Table 8D.5 Stack
contents for the 5-error
case after processing cycle 5.

Inputs	Final encoder state	Path metric
*1000	0	−5.088
*1001	2	−5.088
00	0	−6.792
01	2	−6.792
101	2	−10.188
11	3	−11.04

■ Table 8D.6 Stack
contents for the 5-error
case after processing cycle 6.

Inputs	Final encoder state	Path metric
*10001	2	−4.236
1001	2	−5.088
00	0	−6.792
01	2	−6.792
101	2	−10.188
11	3	−11.04
*10000	0	−12.732

216

■ Table 8D.7 Stack contents for the 5-error case after processing cycle 7.

Inputs	Final encoder state	Path metric
*100011	3	−3.384
1001	2	−5.088
00	0	−6.792
01	2	−6.792
101	2	−10.188
11	3	−11.04
*100010	1	−11.88
10000	0	−12.732

■ Table 8D.8 Stack contents for the 5-error case after processing cycle 8.

Inputs	Final encoder state	Path metric
1001	2	−5.088
*1000110	1	−6.78
*1000111	3	−6.78
00	0	−6.792
01	2	−6.792
101	2	−10.188
11	3	−11.04
100010	1	−11.88
10000	0	−12.732

■ Table 8D.9 Stack contents for the 5-error case after processing cycle 9.

Inputs	Final encoder state	Path metric
1000110	1	−6.78
1000111	3	−6.78
00	0	−6.792
01	2	−6.792
*10010	1	−8.484
*10011	3	−8.484

■ Table 8D.9 Continued.

Inputs	Final encoder state	Path metric
101	2	−10.188
11	3	−11.04
100010	1	−11.88
10000	0	−12.732

■ Table 8D.10 Stack
contents for the 5-error
case after processing cycle 10.

Inputs	Final encoder state	Path metric
*10001100	0	−5.928
1000111	3	−6.78
00	0	−6.792
01	2	−6.792
10010	1	−8.484
10011	3	−8.484
101	2	−10.188
11	3	−11.04
100010	1	−11.88
10000	0	−12.732

■ Table 8D.11 Stack
contents for the 5-error
case after processing cycle 11.

Inputs	Final encoder state	Path metric
1000111	3	−6.78
00	0	−6.792
01	2	−6.792
10010	1	−8.484
10011	3	−8.484
101	2	−10.188
11	3	−11.04
100010	1	−11.88
10000	0	−12.732
*10001100	0	−13.572

Stack Tables for Example 8.3

■ Table 8D.12 Stack contents for the 5-error case after processing cycle 12.

Inputs	Final encoder state	Path metric
00	0	−6.792
01	2	−6.792
10010	1	−8.484
10011	3	−8.484
*10001110	1	−10.176
*10001111	3	−10.176
101	2	−10.188
11	3	−11.04
100010	1	−11.88
10000	0	−12.732
10001100	0	−13.572

■ Table 8D.13 Stack contents for the 5-error case after processing cycle 13.

Inputs	Final encoder state	Path metric
*001	2	−5.94
01	2	−6.792
10010	1	−8.484
10011	3	−8.484
10001110	1	−10.176
10001111	3	−10.176
101	2	−10.188
11	3	−11.04
100010	1	−11.88
10000	0	−12.732
10001100	0	−13.572
*000	0	−14.436

Stack Tables for Example 8.3

■ Table 8D.14 Stack
contents for the 5-error
case after processing cycle 14.

Inputs	Final encoder state	Path metric
*0010	1	−5.088
01	2	−6.792
10010	1	−8.484
10011	3	−8.484
10001110	1	−10.176
10001111	3	−10.176
101	2	−10.188
11	3	−11.04
100010	1	−11.88
10000	0	−12.732
10001100	0	−13.572
*0011	3	−13.584
000	0	−14.436

■ Table 8D.15 Stack
contents for the 5-error
case after processing cycle 15.

Inputs	Final encoder state	Path metric
*00100	0	−4.236
01	2	−6.792
10010	1	−8.484
10011	3	−8.484
10001110	1	−10.176
10001111	3	−10.176
101	2	−10.188
11	3	−11.04
100010	1	−11.88
*00101	2	−12.732
10000	0	−12.732
10001100	0	−13.572
0011	3	−13.584
000	0	−14.436

■ Table 8D.16 Stack
contents for the 5-error
case after processing cycle 16.

Inputs	Final encoder state	Path metric
01	2	−6.792
*001000	0	−7.632
*001001	2	−7.632
10010	1	−8.484
10011	3	−8.484
10001110	1	−10.176
10001111	3	−10.176
101	2	−10.188
11	3	−11.04
100010	1	−11.88
00101	2	−12.732
10000	0	−12.732
10001100	0	−13.572
0011	3	−13.584
000	0	−14.436

■ Table 8D.17 Stack
contents for the 5-error
case after processing cycle 17.

Inputs	Final encoder state	Path metric
001000	0	−7.632
001001	2	−7.632
10010	1	−8.484
10011	3	−8.484
10001110	1	−10.176
10001111	3	−10.176
101	2	−10.188
*010	1	−10.188
*011	3	−10.188
11	3	−11.04
100010	1	−11.88
00101	2	−12.732
10000	0	−12.732
10001100	0	−13.572
0011	3	−13.584
000	0	−14.436

Stack Tables for Example 8.3

■ Table 8D.18 Stack
contents for the 5-error
case after processing cycle 18.

Inputs	Final encoder state	Path metric
*0010001	2	−6.78
001001	2	−7.632
10010	1	−8.484
10011	3	−8.484
10001110	1	−10.176
10001111	3	−10.176
101	2	−10.188
010	1	−10.188
011	3	−10.188
11	3	−11.04
100010	1	−11.88
00101	2	−12.732
10000	0	−12.732
10001100	0	−13.572
0011	3	−13.584
000	0	−14.436
*0010000	0	−15.276

■ Table 8D.19 Stack
contents for the 5-error
case after processing cycle 19.

Inputs	Final encoder state	Path metric
001001	2	−7.632
10010	1	−8.484
10011	3	−8.484
*00100010	1	−10.176
10001110	1	−10.176
10001111	3	−10.176
101	2	−10.188
010	1	−10.188
011	3	−10.188
11	3	−11.04
100010	1	−11.88
00101	2	−12.732
10000	0	−12.732
10001100	0	−13.572
0011	3	−13.584

Stack Tables for Example 8.3

Inputs	Final encoder state	Path metric
000	0	−14.436
0010000	0	−15.276

■ Table 8D.20 Stack
contents for the 5-error
case after processing cycle 20.

Inputs	Final encoder state	Path metric
10010	1	−8.484
10011	3	−8.484
00100010	1	−10.176
10001110	1	−10.176
10001111	3	−10.176
101	2	−10.188
010	1	−10.188
011	3	−10.188
*0010010	1	−11.028
*0010011	3	−11.028
11	3	−11.04
100010	1	−11.88
00101	2	−12.732
10000	0	−12.732
10001100	0	−13.572
0011	3	−13.584
000	0	−14.436
0010000	0	−15.276

■ Table 8D.21 Stack
contents for the 5-error
case after processing cycle 21.

Inputs	Final encoder state	Path metric
10011	3	−8.484
00100010	1	−10.176
10001110	1	−10.176
10001111	3	−10.176
101	2	−10.188

223

Inputs	Final encoder state	Path metric
010	1	−10.188
011	3	−10.188
0010010	1	−11.028
0010011	3	−11.028
11	3	−11.04
*100100	0	−11.88
*100101	2	−11.88
100010	1	−11.88
00101	2	−12.732
10000	0	−12.732
10001100	0	−13.572
0011	3	−13.584
000	0	−14.436
0010000	0	−15.276

■ Table 8D.22 Stack contents for the 5-error case after processing cycle 22.

Inputs	Final encoder state	Path metric
*100110	1	−7.632
00100010	1	−10.176
10001110	1	−10.176
10001111	3	−10.176
101	2	−10.188
010	1	−10.188
011	3	−10.188
0010010	1	−11.028
0010011	3	−11.028
11	3	−11.04
100100	0	−11.88
100101	2	−11.88
100010	1	−11.88
00101	2	−12.732
10000	0	−12.732
10001100	0	−13.572
0011	3	−13.584
000	0	−14.436
0010000	0	−15.276
*100111	3	−16.128

224

Stack Tables for Example 8.3

■ Table 8D.23 Stack contents for the 5-error case after processing cycle 23.

Inputs	Final encoder state	Path metric
*1001100	0	−6.78
00100010	1	−10.176
10001110	1	−10.176
10001111	3	−10.176
101	2	−10.188
010	1	−10.188
011	3	−10.188
0010010	1	−11.028
0010011	3	−11.028
11	3	−11.04
100100	0	−11.88
100101	2	−11.88
100010	1	−11.88
00101	2	−12.732
10000	0	−12.732
10001100	0	−13.572
0011	3	−13.584
000	0	−14.436
0010000	0	−15.276
*1001101	2	−15.276
100111	3	−16.128

■ Table 8D.24 Stack contents for the 5-error case after processing cycle 24.

Inputs	Final encoder state	Path metric
00100010	1	−10.176
10001110	1	−10.176
10001111	3	−10.176
101	2	−10.188
010	1	−10.188
011	3	−10.188
0010010	1	−11.028
0010011	3	−11.028
11	3	−11.04
100100	0	−11.88
100101	2	−11.88

■ Table 8D.24 Continued.

Inputs	Final encoder state	Path metric
100010	1	−11.88
00101	2	−12.732
10000	0	−12.732
10001100	0	−13.572
0011	3	−13.584
*10011000	0	−14.424
000	0	−14.436
0010000	0	−15.276
1001101	2	−15.276
100111	3	−16.128

■ Table 8D.25 Stack contents for the 5-error case after processing cycle 25.

Inputs	Final encoder state	Path metric
*001000100	0	−9.324
10001110	1	−10.176
10001111	3	−10.176
101	2	−10.188
010	1	−10.188
011	3	−10.188
0010010	1	−11.028
0010011	3	−11.028
11	3	−11.04
100100	0	−11.88
100101	2	−11.88
100010	1	−11.88
00101	2	−12.732
10000	0	−12.732
10001100	0	−13.572
0011	3	−13.584
10011000	0	−14.424
000	0	−14.436
0010000	0	−15.276
1001101	2	−15.276
100111	3	−16.128

Stack Tables for Example 8.3

Software

The software provided on the disk is organized into two directories (\bch and \convol), which contain, respectively, software for bch and other block codes and software for convolutional codes. Each of these directories is in turn subdivided into two subdirectories (\flat_c and \c_plus), which contain, respectively, C programs and C++ programs.

C Functions for Operations Involving Galois Fields

The file \bch\flat_c\galfield.c contains a module for creating a Galois field and performing arithmetic within this field. The prototypes for all of the functions in this file are contained in galfield.h, which is provided as Listing E-1. These functions use several support and utility functions, which can be found in the files support.c and utility.c. The following paragraphs provide short description for all of the functions contained in galfield.c.

GaloisField takes no arguments and returns no value. When invoked, this function interacts with the user to either construct a new Galois field and store it to disk, or retrieve a previously stored field from disk. A field should either be created or retrieved before any of the other functions in this module are called.

AddExponFieldElements takes as arguments two field elements in exponential form and returns the sum of these elements also in exponential form.

MultExponFieldElems takes as arguments two field elements in exponential form and returns the product of these elements also in exponential form.

PackedPolyToExp takes as its argument a field element expressed in polynomial form. The coefficients of this polynomial are elements of GF(2), so the polynomial is assumed to be in packed form with one bit per coefficient. The return value is the input field element expressed in exponential form.

GetDegreeOfExtField takes no argument and returns an integer value that equals the degree of the field established by Galois-Field.

DivExponFieldElems takes as its arguments two field elements in exponential form and returns the quotient obtained by dividing the first argument by the second argument. This result is returned in exponential form.

GetNumFieldElems takes no argument and returns an integer value that equals the number of elements in the field established by GaloisField.

GetMinPoly takes as its argument an unsigned integer that equals the exponent for one of the roots of one of the minimal polynomials associated with the field. The return value is the minimal polynomial for which this argument is a root. This return value is in packed form. Note that this function should not be called until after BuildMinimalPolynomials has been called to create the minimal polynomials associated with the current field.

GetDegreeOfMinPoly takes as its argument an unsigned integer that equals the exponent for one of the roots of one of the minimal polynomials associated with the field. The return value is an integer that equals the degree of the minimal polynomial for which the argument is a root.

GetAlphaM takes as its argument an integer between 2 and 12. The return value is a primitive polynomial of degree m solved for α^m. A primitive polynomial in this form is used to construct the extension field GF(2^m).

BuildCyclotomicCosets takes as its arguments two integers. The first of these, p, must be a prime. The second argument is m, where the current field is GF(2^m). BuildCyclotomicCosets partitions the integers $\{1, 2, 3 \dots (p^m - 2)\}$ into cyclotomic cosets and places these cosets in the module variable Coset_Array [].

BuildMinimalPolynomials takes no arguments and returns no value. When invoked, this function causes minimal polynomials to be generated for the current field. Packed form representations of these polynomials are placed in the module variable Minimal_Polynomial [].

BuildOntoPolynomial takes three arguments and returns no value. The first argument is an array containing a polynomial whose co-efficients are elements of GF(2^m). (Unlike minimal polynomials whose coefficients are elements of GF(2), this polynomial cannot be stored in packed form.) The second argument is a pointer to an int that is used to store the degree of the maximum degree term in the polynomial having a nonzero coefficient. The third argument is an element of $GF(2^m)$.

BuildOntoPolynomial multiplies the polynomial contained in the first argument by $(x + r)$ where r is the element specified by the third argument. The resulting polynomial is stored in the array used for the first argument and the degree pointed to by the second argument is updated to reflect the newly increased degree of this polynomial.

BuildExtensionField is the function that is called by GaloisField to create a new field. This function takes two arguments. The first is the degree m of the desired extension field, and the second is a primitive polynomial of degree m solved for α^m. This polynomial is in packed form.

SaveFieldDef is the function that is called by GaloisField to store a field definition on disk.

LoadExistingField is the function that is called by GaloisField to retrieve an existing field definition from disk.

TupleToExpon takes as its argument a field element in m-tuple form and returns the corresponding exponential form.

ExponToTuple takes as its argument a field element in exponential form and returns the corresponding m-tuple form.

Listing E-1

```
#ifndef GALFIELD_H
#define GALFIELD_H

typedef short int exponent;
typedef unsigned short int bits;
typedef unsigned short int element;

void GaloisField(void);

exponent AddExponFieldElems(  exponent addend1,
                              exponent addend2);

exponent MultExponFieldElems( exponent factor1,
                              exponent factor2);

exponent PackedPolyToExp( word element_as_packed_poly);

int GetDegreeOfExtField( void );

exponent DivExponFieldElems(  exponent dividend,
                              exponent divisor);
```

```
unsigned int GetNumFieldElems( void );

word GetMinPoly( unsigned int root_expon);

int GetDegreeOfMinPoly( unsigned int root_expon);

unsigned int GetAlphaM(int m);

void BuildCyclotomicCosets( int p,
                            int m);

void BuildMinimalPolynomials(void);

void BuildOntoPolynomial( element poly_coeff[],
                          int *max_degree,
                          exponent root_as_exp);

void BuildExtensionField( int m, int alpha_m);

void SaveFieldDef(void);

void LoadExistingField(void);

exponent TupleToExpon( element element_as_tuple);

element ExponToTuple( exponent element_as_exponent);

#endif
```

C Functions for Constructing BCH Codes

The file \bch\flat_c\bch.c contains a main program for constructing BCH codes. This program makes use of general purpose and field arithmetic functions contained in galfield.c, polmulsr.c, support.c, and utility.c, as well as a number of functions specific to BCH codes. The prototypes for these specific functions are contained in buildbch.h and cycencd.h, which are shown in Listings E-2 and E-3 respectively. The following paragraphs describe these specific functions.

BuildBchCode takes three arguments—the first specifies the first power of α that is to be included among the roots of the generator polynomial, the second specifies the last power of a that is to be included among the roots of the generator polynomial, and the third specifies the "skip interval" that is to be used between the first power and the last power.

GetDegOfBchGenPoly returns the degree of the generator polynomial for the code created by BuildBchCode.

GetBchGenPoly returns the generator polynomial for the code created by BuildBchCode. This generator polynomial is returned in packed form. The single argument to GetBchGenPoly specifies whether the highest-order term of this polynomial is to be placed on the left or right of the packed word.

Listing E-2

```
#ifndef BUILDBCH_H
#define BUILDBCH_H

void BuildBchCode(  int first_reqd_alpha_power,
                    int last_reqd_alpha_power,
                    int alpha_power_increm);

int GetDegOfBchGenPoly( void );

word GetBchGenPoly( logical high_left );

#endif
```

CyclicEncoder is included with the BCH code functions, but it can actually be used for encoding with any binary cyclic code. This function takes four arguments. The first argument is the code's generator polynomial in packed form. The second argument is the degree of the generator polynomial. The third argument is an array containing the information bits to be encoded. This array must be dimensioned so that there are sufficient empty locations after the information bits to hold the parity bits that will be generated by CyclicEncoder.

Listing E-3

```
#ifndef CYCENCD_H
#define CYCENCD_H

void CyclicEncoder( word gen_poly,
                    int degree_of_gen_poly,
                    unsigned short int code_vector[],
                    int num_info_bits);

#endif
```

C Functions for Decoding BCH Codes

The file \bch\flat_c\decoder.c contains a main program for decoding BCH codes. This program makes use of general-purpose and field arithmetic functions contained in galfield.c, poldivsr.c, polynom.c, support.c, and utility.c, as well as a number of functions specific to decoding BCH codes. The prototypes for these specific functions are contained in berkamp.h, errloc.h, and syndrome.h, which are shown in Listings E-4 through E-6 respectively. The following paragraphs describe these specific functions.

C Functions for Operations Involving Galois Fields

PerformBerlekamp uses the Peterson-Berlekamp method to find the error-locator polynomial from the syndrome of a received BCH code word. This function takes five arguments. The first argument is a vector of field elements in exponential form which represents the syndrome vector. The next third, fourth, and fifth arguments define the specific BCH being used—the third specifies the first power of α included in the code's generator polynomial, the fourth specifies the last power of a, and the fifth specifies the "skip interval." The fifth argument is a vector for holding field elements in exponential form that will be used to return the error-locator polynomial found by PerformBerlekamp.

Listing E-4

```
#ifndef BERKAMP_H
#define BERKAMP_H

#include "galfield.h"

void PerformBerlekamp( exponent syndromeVector[],
                       int firstReqAlphaPower,
                       int lastReqAlphaPower,
                       int alphaPowerIncrem,
                       exponent sigma[MAX_TERMS]);
#endif
```

LocateErrors uses the error-locator polynomial generated by PerformBerlekamp to construct the error vector for the received codeword.

Listing E-5

```
#ifndef ERRLOC_H
#define ERRLOC_H

#include "galfield.h"

void LocateErrors( exponent sigma[],
                   int bit_per_code_word,
                   bits error_vector[]);

#endif
```

ComputeSyndrome, as its name indicates, computes the syndrome for a received codeword. This function takes six arguments. The first argument is an array containing the received codeword, one bit per location. The second argument is the number of bits in the received codeword. The third argument is a vector for holding field elements in exponential form that will be used to return the syndrome vector found by ComputeSyndrome.

Listing E-6

```
#ifndef SYNDROME_H
#define SYNDROME_H

#include "galfield.h"

void ComputeSyndrome( unsigned short int receivedPoly[],
                      int degreeOfReceivedPoly,
                      exponent syndromeVector[],
                      exponent firstAlphaExp,
                      exponent lastAlphaExp,
                      exponent alphaExpIncrem);

#endif
```

C++ Classes for Decoding BCH Codes

The directory \bch\c_plus contains a main program in decoder.cpp as well as a number of C++ classes that can be used for generating and decoding BCH codes. The files support.cpp, poldivsr.cpp, and utility.cpp contain global functions which are not members of any class. Each of the classes is described in the follow paragraphs.

Class BchDecoder is defined in bchdecdr.hpp, which is provided in Listing E-7. This class contains member functions Setup, Compute Syndrome, PerformBerlekamp, and LocateErrors. The operation of these functions is similar to the operation of the like-named C function.

Listing E-7

```
#ifndef SYNDROME_H
#define SYNDROME_H

#include "galfield.h"

void ComputeSyndrome( unsigned short int receivedPoly[],
                      int degreeOfReceivedPoly,
                      exponent syndromeVector[],
                      exponent firstAlphaExp,
                      exponent lastAlphaExp,
                      exponent alphaExpIncrem);

#endif


#ifndef BCHDECDR_HPP
#define BCHDECDR_HPP

#include "galfield.hpp"
#include "polynom.hpp"
#include "binpoly.hpp"

class BchDecoder {
```

```
public:
  BchDecoder();   //constructor
  ~BchDecoder();    //destructor

  void Setup( GaloisField* gf,
              int length_rcvd_sequence,
              exponent first_alpha_expon,
              exponent last_alpha_expon,
              exponent alpha_expon_increm);

  void ComputeSyndrome( BinaryPolynomial* received_poly,
                        exponent syndrome_vector[] );

  void PerformBerlekamp(  PolynomialOverGF* sigma);
  void LocateErrors( bits error_vector[]);
private:
  GaloisField* G_F;
  int Length_Rcvd_Sequence;
  exponent First_Alpha_Expon;
  exponent Last_Alpha_Expon;
  exponent Alpha_Expon_Increm;
  exponent Syndrome_Vector[MAX_BITS];
  PolynomialOverGF* Error_Loc_Poly;
};
#endif
```

Class BinaryPolynomial is defined in binpoly.hpp, which is provided in Listing E-8. This class is used to represent polynomials with binary coefficients and it includes member functions for operating on such polynomials.

Listing E-8

```
#ifndef BINPOLY_HPP
#define BINPOLY_HPP

#include "galfield.hpp"
#include "polynom.hpp"

class BinaryPolynomial {
public:

  BinaryPolynomial(GaloisField* gf)    //constructor
  ~BinaryPolynomial();    //destructor

  void FindDegree(void);

  int GetDegree(void);

  int GetLength(void);

  void Dump(void);

  bits GetCoeff(int n);
```

```
    void TallyWeight( unsigned short int codeVector[],
                      int bitsPerCodeWord,
                      unsigned int weightDistribution[]);
    word FlipPackedPoly( word forward_poly);

    void SetFromPacked( unsigned long int longWord,
                        int first_term,
                        int numb_terms,
                        logical high_left);

    void BinaryPolynomial::SetFromVector( bits input_vector_poly[],
                                          int first_term,
                                          int numb_terms,
                                          logical high_left);
    unsigned short int LongParity(unsigned long int longWord);

    exponent Evaluate( exponent element);

private:
    void Unpack(void);
    void Pack( void);

    GaloisField* G_F;
    bits* Vector_Poly;
    word* Packed_Poly;
    int Numb_Terms;
    int First_Term;
    logical High_Left;
    int Degree_Of_Poly;

};

#endif
```

235

Class PolynomialOverGF is defined in polynom.hpp, which is provided in Listing E-9. This class is used to represent polynomials with coefficients drawn from $GF(2^m)$.

Listing E-9

```
#ifndef POLYNOM_HPP
#define POLYNOM_HPP

#include "galfield.hpp" /* for exponent type */

class PolynomialOverGF {
public:

    PolynomialOverGF(GaloisField* gf);      //constructor
    ~PolynomialOverGF();  //destructor

    void SetToZero(void);
    void SetToUnity(void);

    exponent Evaluate(  exponent element);
```

```
    void AddOnPolynomial( PolynomialOverGF* added_poly );

    void Shift( int num_places_to_shift);

    void MultByFieldElem( exponent element);

    void Copy(  PolynomialOverGF* source_poly);

    exponent GetCoeff( int lambda );

    void Dump( void );

private:
  GaloisField* G_F;
  exponent* Stored_Polynomial;
};

#endif
```

Class GaloisField is defined in galfield.hpp, which is provided in
Listing E-10. This class is used for creating a Galois field and performing arithmetic within this field. This class contains member
functions whose operation is similar to the operation of the like-
named C functions.

Listing E-10

```
#ifndef GALFIELD_HPP
#define GALFIELD_HPP

typedef short int exponent;
typedef unsigned short int bits;
typedef unsigned short int element;

class GaloisField {
public:

  GaloisField();        //constructor
  ~GaloisField();       //destructor

  void InitGaloisField(void);

  exponent AddExponFieldElems(   exponent addend1,
                                 exponent addend2);

  exponent MultExponFieldElems(   exponent factor1,
                                  exponent factor2);

  exponent PackedPolyToExp( word element_as_packed_poly);

  int GetDegreeOfExtField( void );

  exponent DivExponFieldElems(  exponent dividend,
                                exponent divisor);
```

```
    unsigned int GetNumFieldElems( void );

    word GetMinPoly( unsigned int root_expon);

    int GetDegreeOfMinPoly( unsigned int root_expon);

    unsigned int GetAlphaM(int m);

    void BuildCyclotomicCosets( int p,
                                int m);

    void BuildMinimalPolynomials(void);

    void BuildOntoPolynomial( element poly_coeff[],
                              int *max_degree,
                              exponent root_as_exp);

    void BuildExtensionField( int m, int alpha_m);

    void SaveFieldDef(void);

    void LoadExistingField(void);

    exponent TupleToExpon( element element_as_tuple);

    element ExponToTuple( exponent element_as_exponent);

private:

    void PackTuple( bits bit_vector[],
                    unsigned long int *long_word,
                    int num_bits);

    unsigned int Galois_Field[512];
    unsigned int Num_Field_Elements;
    exponent Coset_Array[32][MAX_ELEMENTS];
    word Minimal_Polynomial[MAX_ELEMENTS];
    int Degree_Of_Min_Poly[MAX_ELEMENTS];
    int Num_Cosets;
    int Degree_Of_Field;
    unsigned int Alpha_Module;

};
#endif
```

C Functions for Working with Convolutional Codes

The directory \convo\flat_c contains a number of general-purpose functions for working with convolutional codes. The prototypes for these functions are defined in files mealy.h, metrics.h, and rcvd sig.h, which are provided as Listings E-11, E-12, and E-13, respectively. The following paragraphs describe each of these functions.

MealyEncoder_Init initializes a Mealy machine model of a rate ½ convolutional encoder. Arguments specify the number of shift register stages as well as the two generator polynomials for the code.

MealyEncoder_GetNextState takes as its arguments an integer that specifies the current state and an integer that specifies the current input. The function returns an integer that specifies the corresponding next state.

MealyEncoder_Getoutput takes as its arguments an integer that specifies the current state and an integer that specifies the current input. The function returns an integer that specifies the corresponding output.

MealyEncoder_GetTransitionTrigger takes as its arguments integers that specify the previous state and the current state. The function returns an integer that specifies the input that would have caused a transition from specified previous state to the specified current state.

Listing E-11

```
#ifndef MEALY_H
#define MEALY_H

//   private:
struct MealyEncoder_type {
   int Output_Symbol[4][4];
   int Next_State[4][4];
   int Max_Input;
   int Max_State;
   };
typedef struct MealyEncoder_type MealyEncoder;

// public:

void MealyEncoder_Init( MealyEncoder* this,
                        int num_stages,
                        int gen_poly_1,
                        int gen_poly_2);

int MealyEncoder_GetNextState(  MealyEncoder* this,
                                int current_state,
                                int input);
int MealyEncoder_GetOutput( MealyEncoder* this,
                            int current_state,
                            int input);

int MealyEncoder_GetTransitionTrigger(  MealyEncoder* this,
                                        int previous_state,
                                        int current_state);
#endif
```

MetricTable_Init sets up a table of branch and symbol metrics. An argument specifies the specific type of metric to be generated.

MetricTable_SoftSymbolMetric takes as it arguments the received soft symbol decision and a candidate transmitted symbol. The function returns the symbol metric corresponding to this combination.

MetricTable_GetBranchMetric takes as it arguments the received symbol decision and a candidate transmitted symbol. The function returns the symbol metric corresponding to this combination.

Listing E-12

```
#ifndef MEALY_H
#define MEALY_H

// private:
struct MealyEncoder_type {
  int Output_Symbol[4][4];
  int Next_State[4][4];
  int Max_Input;
  int Max_State;
  };
typedef struct MealyEncoder_type MealyEncoder;

// public:

void MealyEncoder_Init( MealyEncoder* this,
                        int num_stages,
                        int gen_poly_1,
                        int gen_poly_2);

int MealyEncoder_GetNextState( MealyEncoder* this,
                               int current_state,
                               int input);
int MealyEncoder_GetOutput( MealyEncoder* this,
                            int current_state,
                            int input);

int MealyEncoder_GetTransitionTrigger( MealyEncoder* this,
                                       int previous_state,
                                       int current_state);

#endif

#ifndef METRICS_H
#define METRICS_H

//*****************************
//*  Public type information

  enum metric_type_type
    {
    FANO_METRIC,
    HAMMING_METRIC,
    SOFT_METRIC
    };
```

Listing E-12 Continued

```
//   private:

struct MetricTable_type{
  double Metric_Table[8][8];
  enum metric_type_type Metric_Type;
  };
typedef struct MetricTable_type MetricTable;

// public:
//*****************************
//   Public Class Methods

void MetricTable_Init(  MetricTable* this,
                        enum metric_type_type
metric_type);

double MetricTable_SoftSymbolMetric(  MetricTable* this,
                                      int rx_symbol,
                                      int tx_symbol);

double MetricTable_GetBranchMetric( MetricTable* this,
                                    int rx_symbol,
                                    int tx_symbol);

#endif
```

ReceivedSignal_Init initializes the received codeword object.

ReceivedSignal_GetSymbol takes as its argument an integer specifying a level in the decoder trellis and returns the received symbol for this level.

ReceivedSignal_GetRxLength returns the length (in symbols) of the currently defined received codeword.

Listing E-13

```
#ifndef MEALY_H
#define MEALY_H

//   private:
struct MealyEncoder_type {
  int Output_Symbol[4][4];
  int Next_State[4][4];
  int Max_Input;
  int Max_State;
  };
typedef struct MealyEncoder_type MealyEncoder;

// public:

void MealyEncoder_Init( MealyEncoder* this,
                        int num_stages,
                        int gen_poly_1,
                        int gen_poly_2);
```

```
int MealyEncoder_GetNextState(  MealyEncoder* this,
                                int current_state,
                                int input);
int MealyEncoder_GetOutput( MealyEncoder* this,
                            int current_state,
                            int input);

int MealyEncoder_GetTransitionTrigger(  MealyEncoder* this,
                                        int previous_state,
                                        int current_state);

#endif

#ifndef RCVDSIG_H
#define RCVDSIG_H

#include "metrics.h"

// private:

struct ReceivedSignal_type{
  int Rcvd_Sequence[20];
  int Rx_Length;
  };
typedef struct ReceivedSignal_type ReceivedSignal;

//  public:

//*******************************
//  Public Class Methods

void ReceivedSignal_Init( ReceivedSignal* this,
                          enum metric_type_type metric_type);

int ReceivedSignal_GetSymbol( ReceivedSignal* this,
                              int level);

int ReceivedSignal_GetRxLength(ReceivedSignal* this );

#endif
```

241

C++ Classes for Working with Convolutional Codes

The directory \convol\c_plus contains a number of general-purpose classes for working with convolutional codes. The prototypes for these functions are defined in files mealy.hpp, metrics.hpp, and rcvdsig.hpp. The organization and functionality of these classes is very similar to the organization and functionality of the modules described in the section "C Functions for Working with Convolutional Codes." In fact, the C++ classes were written first, and then the C modules were obtained via rote translation from C++ to C.

C Functions for Viterbi Decoding

The directory \convol\flat_c contains a main program in vitmain.cpp as well as two C functions for Viterbi decoding of convolutional codes. Prototypes for these functions are defined in vitalg.h, which is provided as Listing E-14.

Listing E-14

```
#ifndef VITALG_H
#define VITALG_H

//private:
struct ViterbiDecoder_type{
  int Info_Length;
  int Tail_Length;
  int Frame_Length;
  int Numb_Encdr_States;
  int Numb_Inp_Symbs;
  MealyEncoder* Tx_Encoder;
  MetricTable* Decoding_Metric;
  };
typedef struct ViterbiDecoder_type ViterbiDecoder;

//public:

  void ViterbiDecoder_Init( ViterbiDecoder* this,
                            int info_length,
                            int tail_length,
                            MealyEncoder* encoder,
                            MetricTable* decoding_metric);

  void ViterbiDecoder_Decode( ViterbiDecoder* this,
                              ReceivedSignal* rcvd_signal);


#endif
```

C++ Classes for Viterbi Decoding

The directory \convol\c_plus contains a main program in vitmain .cpp, as well as two C functions for Viterbi decoding of convolutional codes. Prototypes for these functions are defined in vitalg .hpp, which is provided as Listing E-15.

Listing E-15

```
#ifndef VITALG_H
#define VITALG_H

class ViterbiDecoder {
public:
  ViterbiDecoder();
```

```
~ViterbiDecoder();

void Init(  int info_length,
            int tail_length,
            MealyEncoder* encoder,
            MetricTable* decoding_metric);

void Decode(  ReceivedSignal* rcvd_signal);

private:
  int Info_Length;
  int Tail_Length;
  int Frame_Length;
  int Numb_Encdr_States;
  int Numb_Inp_Symbs;
  MealyEncoder* Tx_Encoder;
  MetricTable* Decoding_Metric;
};

#endif
```

C Functions for Stack Decoding

The directory \convol\flat_c contains a main program in stacmain.c, as well as a number of C functions for stack decoding of convolutional codes. Prototypes for these functions are defined in stack.h, stackalg.h, and stackent.h.

The file stack.h includes definitions for a generic stack object and for functions to perform operations upon this stack. This stack is constructed out of stack entry objects, which are defined in stackent.h.

The file stackalg.h includes definitions for a stack algorithm object and for methods that initialize this object and use it to perform stack decoding, as described in Sections 8.1 and 8.2.

C++ Classes for Stack Decoding

The directory \convol\c_plus contains a main program in stacmain .cpp, as well as a number of C classes for stack decoding of convolutional codes. These classes are defined in stack.hpp, stackalg .hpp, and stackent.hpp.

C Functions for Fano Decoding

The directory \convol\flat_c contains a main program in fanomain.c, as well as a number of C functions for using Fano's algorithm to decode convolutional codes. Prototypes for these functions are defined in fanoalg.h, lookfwrd.h, and path.h.

The file fanoalg.h defines data structures for a Fano algorithm object and function prototypes for methods that initialize this object and use it to perform decoding as described in Section 8.3.

The file lookfwrd.h defines data structures and methods for performing the "the look forward" step of the Fano algorithm as described in Section 8.3.

C++ Classes for Fano Decoding

The directory \convol\c_plus contains a main program in fanomain .cpp, as well as a number of C functions for using Fano's algorithm to decode convolutional codes. These classes are defined in fanoalg .hpp, lookfwrd.hpp, and path.hpp.

Index

A

abelian groups, 10, 12
abstract algebra, ix
additive Gaussian noise errors, 1
additive operations, 18
AND operation in Boolean algebra, 7, **7**
associative operatives, 9, 11, 12
augmented cyclic codes, 55
authentication techniques, 1

B

BCH codes, ix, 33, 61-84
  algorithmic approach, 71-73
  binary coefficients, 63
  binding, BCH bound, 61-62
  BuildBchCode( ) program listing, 78-79
  ComputeSyndrome( ) program listing, 96, 100
  critical features of BCH codes, 63-67
  cyclotomic cosets, 33-36, 65-66
  design distances, 62, 68
  error location using Peterson-Berlekamp method, 102-103
  extension fields, 63, 80, **80**
  generating BCH codes, 62-63
  generator polynomials, 68-71, 71-73, 73-79
  isomorphs, 71
  LocateErrors( ) program listing, 102-103
  manual encoding for cyclic codes used, 69
  narrow-sense BCH codes, 63, 64, 66, **66**, 68-71, 71-73, 73-79
  nonbinary BCH codes, 79-82
  nonprimitive BCH codes, 63, 64
  PerformBerlekamp( ) program listing, 100-102
  Peterson-Berlekamp encoding and decoding, 97-103
  prime fields, 63
  primitive BCH codes, 63, 66, **66**, 68-71, 71-73, 73-79
  Reed-Solomon codes, 82-84, **84**
  syndromes, 93-96
  wide-sense BCH codes, 63
binary erasure channel (BEC) errors, 2, **3**
binary extension fields
  creating binary extension fields, 22-24, **24**
  irreducibile polynomials, 21
  primitive polynomials, 20-21, 22
  representing binary extension field elements, 24-25
binary multiplication, 7, **8**
binary symmetric channel (BSC) errors, 1, **2**
binding, BCH bound, 61-62
bit-by-bit modulo-2 arithmetic, 18-19, **18**, **19**
block codes, ix, 41-49, **42**, **43**
  augmented cyclic codes, 55
  code word, code vector, code block, 41
  cyclic codes (*see* cyclic codes)
  dual code, 49
  errors in block codes, 41-42
  expurgated cyclic codes, 54-55, **55**
  extended cyclic codes, 53, **54**
  generator matrices, 44-45, 47
  Hamming codes, 47, 56-59
  how it works, 43-44
  lengthened cyclic codes, 56
  linear block codes, 44-49
  message blocks, 41
  parity, 43-44
  parity-check matrix, 47-49
  programming, 45-47

Illustrations are indicated by **boldface.**

block codes *continued*
 punctured cyclic codes, 54
 shortened cyclic codes, 56, **56**
 shortened Hamming codes, 58-50, **59**
 standard array encoding and decoding,
  89-93, **91**
 syndromes, 48
 systematic vs. nonsystematic code,
  45, **46**, **47**
Boolean algebra, 6, 7, **7**, 8-9
Bose-Chaudhuri-Hocquenghem (*see* BCH
 codes)
branch metrics, Viterbi decoding, 129
BuildBchCode( ) program listing, 78-79
burst channel errors, 1

C

C and C++ programming languages, 2
Clark and Cain convention for convolutional
 codes, 114, 115-116
closed operations, 9, 12
code rates, convolutional codes, 116
code word, code vector, code block, 41
coding, 1
coefficients, polynomial arithmetic, 30-32
commutative groups, 10, 12, 13, 18
compression, data, 1
ComputeSyndrome( ) program listing, 96,
 100
constraint lengths, convolutional codes, 114,
 115
convolutional codes, ix, 105-125
 canonical examples, 105-112, **106**, 127,
  128, 129
 Clark and Cain convention, 114, 115-116
 code rates, 116
 constraint lengths, 114, 115
 conventions, 112-116
 cycles, 114-115
 distance measures, 123-125
 Fano decoding, 161-168, **165**, **167**, **168**,
  169, 180-189, **181**
 fractional rate loss, 116
 homogeneous vs. nonhomogeneous states,
  109-110, **110**, **111**
 how it works, 105-112
 level outputs of Moore machine, 112
 Lin and Costello convention, 113-114,
  115, 116
 Mealy machines, 106-109, **109**, 109-112,
  110-113

Michelson and Levesque convention,
  114-115
 minimum free distance, 124-125, 146-147
 Moore machines, 106, **107**, 109-112
 notation and terminology, 112-116
 partial trellis by collapsing tree, 133-135,
  134, **135-141**
 pulse output of Mealy machine, 112
 sequential decoding, 155-189
 slings, 111
 splitting state of Mealy machine, 111-112,
  112, **113**
 stack decoding, 155-161, **157**, **159**, **160**
 tree representation , 116-119, **120**, **121**,
  122, 129-131, **129**, **131**, **133**
 trellis representation, 119, 123, **123**,
  124, 127, **128**, 129, **142-146**
 truncated code rates, 116
 Viterbi convention, 115
 Viterbi decoding, 127-153
 weight distribution errors, **145**, 147
coset leader, standard array encoding and
 decoding, 90
cryptographic coding, 1
cycles, convolutional codes, 114-115
cyclic codes, ix, 49-56, **50**
 augmented cyclic codes, 55
 codewords showing cyclical shift, **50**
 CyclicEncoder( ) program listing, 88-89
 division method encoding and decoding,
  85-89, **86**, **87**, **88**
 expurgated cyclic codes, 54-55, **55**
 extended cyclic codes, 53, **54**
 generator polynomials, 51
 lengthened cyclic codes, 56
 manual encoding methods, 51-52, 69
 punctured cyclic codes, 54
 shortened cyclic codes, 56, **56**
CyclicEncoder( ) program listing, 88-89
cyclotomic cosets, 33-36
 BCH codes, 65-66

D

data compression, 1
data errors (*see* errors, data errors)
deadband errors, 2
decoding (*see* encoding and decoding)
degrees, polynomial, 14
design distances, BCH codes, 62, 68
discrete memory-less channel (DMC) errors, 1
 Viterbi decoding, 148, **149**

246

distance measures, convolutional codes, 123-125

distributive operations, 12

division method encoding and decoding, cyclic codes, 85-89, **86**, **87**, **88**

dual code, block codes, 49

E

encoding and decoding, 85-103

  BCH codes

    Peterson-Berlekamp encoding and decoding, 97-10

    syndromes, 93-96

  block codes, standard array encoding and decoding, 89-93

  ComputeSyndrome() program listing, 96, 100

  convolutional codes (*see* convolutional codes)

  cyclic codes, division method encoding and decoding, 85-89

  CyclicEncoder() program listing, 88-89

  division method encoding and decoding, 85-89, **86**, **87**, 88

  error location using Peterson-Berlekamp method, 102-103

  Fano decoding, 155, 161-168, **165**, **167**, **168**, **169**, 180-189, **181**

  LocateErrors() program listing, 102-103

  PerformBerlekamp() program listing, 100-102

  Peterson-Berlekamp encoding and decoding, BCH codes, 97-10

  sequential decoding (*see* sequential decoding)

  stack decoding, 155-161, **157**, **159**, **160**

  standard array encoding and decoding, 89-93, **91**

  syndromes, BCH codes, 93-96

  Viterbi decoding (*see* Viterbi decoding)

erasure errors, 2

error control codes, 1

error detection and correction (EDAC) codes, 5

errors, data errors, 1-2

  additive Gaussian noise errors, 1

  binary erasure channel (BEC) errors, 2, **3**

  binary symmetric channel (BSC) errors, 1, **2**

  block codes, 41-42

  burst channel errors, 1

  channel-with-memory errors, 1

  deadband errors, 2

  discrete memory-less channel (DMC) errors, 1, 148, **149**

  erasure errors, 2

  noise, 1

expurgated cyclic codes, 54-55, **55**

extended cyclic codes, 53, **54**

extension fields, 5-6, 16-20

  BCH codes, 63, 80, **80**

  binary extension fields, 20-25

  creating fields, 17-20

  ground fields, 19

  modulo-8 arithmetic, 16-17, **17**

  polynomial arithmetic, 17

  prime fields, 19

F

Fano decoding, ix, 155, 161-168, **165**, **167**, **168**, **169**, 180-189, **181**

finite (Galois) fields, ix, 9-13, 26-30

  abelian groups, 10, 12

  associative operatives, 9, 11, 12

  brute force approach to field creation, 28

  closed operations, 9, 12

  commutative groups, 10, 12, 13

  distributive operations, 12

  fields defined, 11-13

  finite (Galois) fields, 9, 12, 13

  groups, 9

  identity elements, 9, 11, 12

  infinite fields, 9

  inverse operations, 9, 12

  modulo-3 arithmetic, 10-11, **10**

  modulo-4 arithmetic, 11, **11**

  order of fields, 12

  prime fields, 12, 13

  programming, 26-30

  representing field elements, 27-29

  table sizes and iteration counts, 28, **29**

fractional rate loss, convolutional codes, 116

G

Galois fields (*see* finite (Galois) fields)

Gaussian noise errors, 1

generator matrices, block codes, 44-45, 47

generator polynomials

  BCH codes, 68-71, 71-73, 73-79

  cyclic codes, 51

global variables in programming, 3
ground fields, 19
groups, 9

H

Hamming codes, 47
homogeneous vs. nonhomogeneous states,
    Mealy machines, 109-110, **110**, **111**

I

identity elements, 9, 11, 12, 18
inverse operations, 9, 12, 18
irreducibile polynomials, 21
isomorphs, 7, 71

L

lengthened cyclic codes, 56
Lin and Costello convention for
    convolutional codes, 113-114, **115**, 116
linear block codes, 44-49
linked-list implementation of stack decoding,
    172-174, **172**, **173**, **174**
LocateErrors( ) program listing, 102-103
logical operations in Boolean algebra, 7, **7**, 8-
    9

M

math coding tools, 5-40
    binary extension fields, 20-25
    Boolean algebra, 6, 7, **7**, 8-9
    cyclotomic cosets, 33-36
    extension fields, 5, 16-20
    finite (Galois) fields, 9-13, 26-30
    modulo-p arithmetic, 5-9
    polynomial arithmetic, 5, 13-16, 17, 30-32,
        36-40
Mealy machines
    C functions, 231-234
    convolutional codes, 106-109, **109**,
        109-112, 110, **111**, **112**, **113**
message blocks, 41
Michelson and Levesque convention for
    convolutional codes, 114-115
minimal polynomials, 36-40
minimum free distance, convolutional codes,
    124-125, 146-147
modular programming, 2-3
modulo-2 arithmetic, 6, 7-9, **8**, 18, **18**

modulo-3 arithmetic, 10-11, **10**
modulo-4 arithmetic, 11, **11**
modulo-5 arithmetic, 6, **6**
modulo-8 arithmetic, 16-17, **17**
modulo-p arithmetic, 5-9
    abelian groups, 10, 12
    additive operations, 18
    AND operation in Boolean algebra, 7, **7**
    associative operatives, 9, 11, 12
    binary multiplication, 7, **8**
    bit-by-bit modulo-2 arithmetic, 18-19, **18**,
        19
    Boolean algebra, 6, 7, **7**
    closed operations, 9, 12
    commutative groups, 10, 12, 13, 18
    creating fields, 17-20
    distributive operations, 12
    extension fields, 6, 16-20, 19
    fields, 5-6
    finite (Galois) fields, 12, 13
    ground fields, 19
    groups, 9
    identity elements, 9, 11, 12, 18
    inverse operations, 9, 12, 18
    isomorphism in Boolean algebra, 7
    logical operations in Boolean algebra,
        7, **7**, 8-9
    modulo-2 arithmetic, 6, 7-9, **8**, 18, **18**
    modulo-3 arithmetic, 10-11, **10**
    modulo-4 arithmetic, 11, **11**
    modulo-5 arithmetic, 6, **6**
    modulo-8 arithmetic, 16-17, **17**
    number system for modulo-5 arithmetic,
        6, **6**
    order of fields, 12
    prime fields, 6, 12, 13, 19
    programming for modulo-p arithmetic, 8-9
    remainder operator in modulo-p
        arithmetic, 8
    True/False values in Boolean algebra, 7
    XOR operation in Boolean algebra, 7, **7**
Moore machines, convolutional codes, 106,
    107, 109-112

N

narrow-sense BCH codes, 63, 64, 66, **66**, 68-
    71, 71-73, 73-79
noise, 1
nonbinary BCH codes, 79-82
nonprimitive BCH codes, 63, 64
number system for modulo-5 arithmetic, 6, **6**

O

object-based programming (OOP), 2-3
order of fields, 12
order, polynomial order, 14

P

parity, block codes, 43-44
parity-check matrix, block codes, 47-49
path metrics, Viterbi decoding, 130
PerformBerlekamp( ) program listing, 100-102
Peterson-Berlekamp encoding and decoding, BCH codes, 97-103
  error location, 102-103
polynomial arithmetic, 5, 13-16, 17, 30-32, 36-40
  binary extension fields, 20-25
  coefficient values, 30-32
  creating binary extension fields, 22-24, **24**
  cyclotomic cosets, 33-36
  degrees, polynomial, 14
  exponential forms, 31-32
  extension fields, 19
  ground fields, 19
  irreducibile polynomials, 21
  minimal polynomials, 36-40
  notation for polynomials, 14
  order, polynomial order, 14
  prime fields, 19
  primitive polynomials, 20-21, 22
  programming, 32, 38-40
  programming polynomials, 15
  representing binary extension field elements, 24-25
  weighting, 13
prime fields, 6, 12, 13, 19
  BCH codes, 63
primitive BCH codes, 63, 64, 65-66, **66**, 68-71, 71-73, 73-79
primitive polynomials, 20-21, 22
programming basics, 2-3
  BCH codes, 73-79, **76**
  block codes, 45-47
  C and C++ programming languages, 2, 224-236
  cyclotomic cosets, 35-36
  finite (Galois) fields, 26-30
  minimal polynomials, 38-40
  modular programming, 2-3
  modulo-p arithmetic, 8-9

object-based programming (OOP), 2-3
  polynomial arithmetic, 32
  polynomials, 15-16, 15
  variables, static and global, 3
punctured cyclic codes, 54

R

received symbol buffering, stack decoding, 177-179, **178**
Reed-Solomon codes, ix, 33, 61, 82-84, **84**
  cyclotomic cosets, 33-36
remainder operator in modulo-p arithmetic, 8

S

sequential decoding, 155-189
  Fano decoding, 155, 161-168, **165**, **167**, **168**, **169**, 180-189, **181**
  stack decoding, 155-161, **157**, **159**, **160**, 168-180
shortened cyclic codes, 56, **56**
shortened Hamming codes, 58-50, **59**
slings, convolutional codes, 111
soft decisions and Viterbi decoding, 147-152
splitting state of Mealy machine, 111-112, **112**, **113**
stack decoding, ix, 155-161, **157**, **159**, **160**, 168-180
  buffers, stack buffers, 172, 174-177, **175**, 177-179
  elements of stack, 171
  Fano decoding, 155, 161-168, **165**, **167**, **168**, **169**, 180-189, **181**
  implementing the stack, 171
  linked-list implementation of stack, 172-174, **172**, **173**, **174**
  received symbol buffering, 177-179, **178**
  software approach, 168-180
  stack implementation of stack, 174-177, **175**
  state tables, 179, **180**
  structure of stack, 171-172
  tables, stack table examples, 191-220
  tables, stack tables, 171
standard array encoding and decoding, block codes, 89-93, **91**
static variables in programming, 3
syndromes
  BCH codes, 93-96
  block codes, 48
systematic vs. nonsystematic code, block codes, 45, **46**, **47**

249

T

tree representation, Viterbi decoding, 129-131, **129**, **131**, **133**

tree representation, convolutional codes, 116-119, **120**, **121**, **122**

trellis representation, convolutional codes, 119, 123, **123**, **124**, 127, **128**, 129, **142-146**

True/False values in Boolean algebra, 7

truncated code rates, convolutional codes, 116

V

variables, static and global, 3

Viterbi convention for convolutional codes, 115

Viterbi decoding, ix, 127-153
 branch metrics, 129
 depth of decoding, 153
 discrete memoryless channel (DMC), 148, **149**
 failures of Viterbi decoding, 141-147
 hard decisions, 148
 how it works, 127-141
 minimum free distance, 146-147
 partial trellis by collapsing tree, 133-135, **134**, **135-141**
 path metrics, 130
 practical considerations, 152-153
 soft decisions and Viterbi decoding, 147-152
 tree representation, Viterbi decoding, 129-131, **129**, **131**, **133**
 trellis representation, convolutional codes, 127, **128**, 129, **142-146**
 weight distribution errors, **145**, 147

W

weighting, 13

wide-sense BCH codes, 63

X

XOR operation in Boolean algebra, 7, **7**

About the author

C. Britton Rorabaugh is a working engineer who holds BSEE and MSEE degrees from Drexel University. His previous books include *Digital Filter Designer's Handbook*, *Circuit Design and Analysis*, and *Communications Formulas and Algorithms*, published by McGraw-Hill, and *Data Communications and LAN Handbook* and *Signal Processing Design Techniques*, published by TAB Professional and Reference Books, an imprint of McGraw-Hill.

Other Bestsellers of Related Interest

Single Sideband Systems and Circuits, 2nd Edition
—William E. Sabin and Edgar O. Schoenike

Written for communications, radio, and antenna engineers, this reference covers the use of computer algorithms in SSB design, new hardware and software developments, and a case study of a state-of-the-art system. It includes system design considerations, receiver design, exciter and transceiver design, algorithms, digital signal processing, preselectors and postselectors, synthesizers for SSB, frequency standards, amplifiers, power supplies, antenna matching techniques, and more.

0-07-912038-5 $75.00 Hardcover/Disk

**Mobile Communications Satellites:
Theory and Applications**
—Tom Logsdon

This pioneering book details the present status and future potential of mobile communication satellite constellations currently being deployed in geosynchronous, low-altitude, and medium-altitude orbits. It includes discussion of how the most efficient constellation architectures are selected using analysis tools, how the satellites are designed, what signal formats and frequencies they use, and how they are being applied today to revolutionize communications between mobile systems around the world.

0-07-038476-2 $50.00 Hardcover

Satellite Communication Systems: Design Principles
—M. Richharia

This book provides an in-depth look at the basics of designing the elements and components that comprise a satellite communications system, including the geosynchronous satellites prevalent in mobile communications. Aimed at a broad spectrum of readers—from engineering professionals to graduate students—the book focuses on design fundamentals rather than specific systems, and eschews long mathematical derivations in favor of practical design concepts, guidelines, and models.

0-07-052374-6 $55.00 Hardcover

Signal and Power Integrity in Digital Systems: TTL, CMOS, and BiCMOS
—James E. Buchanan

Here is the first book to show readers how to ensure signal integrity and control noise during the design of high speed logic devices. Spotlighting TTL, CMOS, and BiCMOS, the book covers a wide variety of relevant interconnection and timing issues. Readers will also appreciate the practical engineering approximations provided for the calculation of design parameters.

0-07-008734-2 $60.00 Hardcover

Analog to Digital Conversion—A Practical Approach
—Kevin M. Dougherty

Save money with this comprehensive and practical guide that gives you low-cost techniques and design requirements for high-performance solutions. Hardware and software techniques are fully explored along with support components and layout practices. Engineers faced with the conversion task will be able to quickly choose an optimum approach and understand how to make it work with the easy-to-use information and clear examples.

0-07-015675-1 $55.00 Hardcover

The Economics of Automatic Testing, 2nd Edition
—Brendan Davis

Davis shows quality, manufacturing, and test engineers how to develop an optimum test strategy and sell it to management amid today's often conflicting pressures to cut costs, use the latest technology, and bring new products to market in record time. The Economics of Automatic Testing helps companies weigh such factors and come to a rational, cost-justified decision on whether and how to implement automatic testing.

0-07-707792-X $60.00 Hardcover

DISK WARRANTY

This software is protected by both United States copyright law and international copyright treaty provision. You must treat this software just like a book, except that you may copy it into a computer in order to be used and you may make archival copies of the software for the sole purpose of backing up our software and protecting your investment from loss.

By saying "just like a book," McGraw-Hill means, for example, that this software may be used by any number of people and may be freely moved from one computer location to another, so long as there is no possibility of its being used at one location or on one computer while it also is being used at another. Just as a book cannot be read by two different people in two different places at the same time, neither can the software be used by two different people in two different places at the same time (unless, of course, McGraw-Hill's copyright is being violated).

LIMITED WARRANTY

McGraw-Hill takes great care to provide you with top-quality software, thoroughly checked to prevent virus infections. McGraw-Hill warrants the physical diskette(s) contained herein to be free of defects in materials and workmanship for a period of sixty days from the purchase date. If McGraw-Hill receives written notification within the warranty period of defects in materials or workmanship, and such notification is determined by McGraw-Hill to be correct, McGraw-Hill will replace the defective diskette(s). Send requests to:

> McGraw-Hill
> Customer Services
> P.O. Box 545
> Blacklick, OH 43004-0545

The entire and exclusive liability and remedy for breach of this Limited Warranty shall be limited to replacement of defective diskette(s) and shall not include or extend to any claim for or right to cover any other damages, including but not limited to, loss of profit, data, or use of the software, or special, incidental, or consequential damages or other similar claims, even if McGraw-Hill has been specifically advised of the possibility of such damages. In no event will McGraw-Hill's liability for any damages to you or any other person ever exceed the lower of suggested list price or actual price paid for the license to use the software, regardless of any form of the claim.

McGRAW-HILL SPECIFICALLY DISCLAIMS ALL OTHER WARRANTIES, EXPRESS OR IMPLIED, INCLUDING, BUT NOT LIMITED TO, ANY IMPLIED WARRANTY OF MERCHANTABILITY OR FITNESS FOR A PARTICULAR PURPOSE.

Specifically, McGraw-Hill makes no representation or warranty that the software is fit for any particular purpose and any implied warranty of merchantability is limited to the sixty-day duration of the Limited Warranty covering the physical diskette(s) only (and not the software) and is otherwise expressly and specifically disclaimed.

This limited warranty gives you specific legal rights; you may have others which may vary from state to state. Some states do not allow the exclusion of incidental or consequential damages, or the limitation on how long an implied warranty lasts, so some of the above may not apply to you.